T0135713

Metal-Catalyzed N–H and O–H Insertion from α-Diazocarbonyl Compounds

Zur Erlangung des akademischen Grades eines

DOKTORS DER NATURWISSENSCHAFTEN

(Dr. rer. nat.)

von der KIT-Fakultät für Chemie und Biowissenschaften
des Karlsruher Instituts für Technologie (KIT)

genehmigte

DISSERTATION

von

Yuling Hu M.Sc.
aus Shaanxi, China

Dekan: Prof. Dr. Reinhard Fischer

Referent: Prof. Dr. Stefan Bräse

Korreferent: Prof. Dr. Joachim Podlech

Tag der mündlichen Prüfung: 16.04.2018

Band 79
Beiträge zur organischen Synthese
Hrsg.: Stefan Bräse

Prof. Dr. Stefan Bräse
Institut für Organische Chemie
Karlsruher Institut für Technologie (KIT)
Fritz-Haber-Weg 6
D-76131 Karlsruhe

Bibliografische Information der Deutschen Bibliothek

Die Deutsche Nationalbibliothek verzeichnet diese Publikation in der
Deutschen Nationalbibliografie; detaillierte bibliografische Daten sind
im Internet über http://dnb.d-nb.de abrufbar.

ISBN 978-3-8325-4864-3
ISSN 1862-5681

Logos Verlag Berlin GmbH
Comeniushof, Gubener Str. 47,
10243 Berlin
Tel.: +49 030 42 85 10 90
Fax: +49 030 42 85 10 92
INTERNET: http://www.logos-verlag.de

Fortune always appreciates a hardworking man.

勤能补拙

Chinese proverb

Die vorliegende Arbeit wurde in der Zeit von September 2014 bis März 2018 am Institut für Organische Chemie, der Fakultät für Chemie und Biowissenschaften am Karlsruher Institut für Technologie (KIT) unter der Leitung von Prof. Dr. Stefan Bräse angefertigt.

Die Arbeit wurde durch ein Promotionsstipendium des Chinese Scholarship Council (CSC) gefördert.

Hiermit versichere ich, die vorliegende Arbeit selbstständig verfasst und keine anderen als die angegebenen Quellen und Hilfsmittel verwendet sowie Zitate kenntlich gemacht zu haben. Die Dissertation wurde bisher an keiner anderen Hochschule oder Universität eingereicht.

Contents

1 Abstract

α-Diazocarbonyl compounds are useful and versatile synthons in organic chemistry. They can be prepared from readily accessible precursors and undergo a wide range of chemical transformation reactions. Among them, the transition-metal catalyzed X–H insertions (X = C, N, O, S, B and Si) *via* the decomposition of α-diazocarbonyl compounds to *in situ* generate metal carbenes or carbenoids, represent one of the most efficient approaches to form carbon-heteroatom bonds, which are ubiquitous and among the most reactive components of organic compounds. The presented thesis is based on the synthesis and application of α-diazocarbonyl compounds in metal-catalyzed N–H insertion reactions and the chemoenzymatic synthesis of heterocycles *via* a [Cu]/[Rh] catalyzed intramolecular O–H insertion.

The first part of this work is focused on the synthesis of eleven saturated α-diazocarbonyl compounds *via* the diazo transfer method in up to 95% yield. A straightforward approach *via* the WITTIG reaction, γ-Umpolung addition and diazo transfer reaction allowed the synthesis of unsaturated α-diazocarbonyl compounds with ease. Potassium fluoride was proven to be the optimal base in diazo transfer procedure due to the facile deprotonation of β-acetoacetates by its 'naked' anion of the fluoride.

In the second part, metal-catalyzed N–H insertion with [2.2]paracyclophane-based ligands are investigated. The newly synthesized (R_p)-pseudo-*ortho* [2.2]paracyclophane-based bisoxazoline ligands presented superior activity in comparison to (S,S)-Ph-Pybox. The obtained δ-amino α,β-unsaturated carboxylic esters were used to synthesize hexahydroindoles, which are key intermediates in the synthesis route of Rostratin B–D.

Additional investigation of α-diazocarbonyl compounds led to the development of enzyme catalyzed highly enantioselective reductions. The ketoreductases LbADH and Gre3P performed excellent enantioselectivity to generate (R)- and (S)-hydroxyl α-diazocarbonyl compounds, respectively. The stability of the diazo group during the reaction provides an efficient method to access the chiral alcohol with the retention of the diazo group for further functionalization. The hydroxyl α-diazocarbonyl compounds were used in metal-catalyzed intramolecular O–H insertions. Yields of up to 81% were obtained when $Rh_2(OAc)_4$ was used as a catalyst while $CuOTf \cdot Tol_{1/2}$ resulted in 7:93 diastereoselectivity.

Kurzzusammenfassung – Abstract in German

α-Diazocarbonyle sind nützliche und vielseitige Bausteine in der organischen Chemie. Man kann sie ausgehend von verfügbaren Vorläufermolekülen synthetisieren und sie können eine Vielfalt an chemischen Reaktionen eingehen. Übergangsmetall-katalysierte X–H Insertionen (X = C, N, O, S, B und Si) *via* Spaltung der α-Diazocarbonyle, um *in situ* Metallcarbene oder–carbenoide zu generieren, repräsentieren einen der effizientesten Ansätze, um Heteroatom-Kohlenstoff-Bindungen zu knüpfen. Diese sind ubiquitär und gehören zu den reaktivsten Komponenten organischer Moleküle. Die vorliegende Arbeit beschäftigt sich mit der Synthese und Anwendung von α-Diazocarbonylen in metallkatalysierten X–H Insertionen und der chemoenzymatischen Synthese von Heterozyklen *via* einer [Cu]/[Rh] katalysierten intramolekularen O–H Insertion.

Im ersten Teil der Arbeit wurden elf gesättigte α-Diazocarbonyle mittels Diazo-Transfer in bis zu 95% Ausbeute synthetisiert. Ein direkter Zugang über die WITTIG Reaktion, γ-Umpolung Addition und Diazo-Transfer ermöglicht die simple Synthese ungesättigter α-Diazocarbonyle. Es zeigte sich, dass Kaliumfluorid die beste Base für den Diazo-Transfer darstellt, da Kaliumfluorid die einfache Deprotonierung von β-Acetoacetaten durch das freie Anion des Fluorids ermöglicht.

Im zweiten Teil der Arbeit wurde die metallkatalysierte N–H Insertion mit [2.2]Paracyclophan Liganden erforscht. Der neu dargestellte (R_p)-pseudo-*ortho* [2.2]Paracyclophan-Bisoxazolin Ligand zeigte starke Aktivität im Vergleich zu Ruthenium-Katalysatoren, die auf einem [2.2]Paracyclophan Rückgrat basieren. Die erhaltenen δ-Amino α,β-ungesättigten Carboxylester wurden verwendet, um Hexahydroindole zu synthetisieren, die Schlüsselintermediate in der Synthese von Rostratin B–D darstellen.

Die Arbeit mit α-Diazocarbonylen führte zur Entwicklung enzymkatalysierter stark enantioselektiver Reduktionen. Zum Beispiel arbeiten die Ketoreduktasen LbADH und Gre3P bei der Synthese von (R)- und (S)-Hydroxyl α-Diazocarbonyl-Verbindungen sehr enantioselektiv. Die hohe Stabilität der Diazogruppen in der Reaktion führt zu einer effizienten Methode zur Darstellung des chiralen Alkohols unter Retention des Amins, welches wiederum für weitere Funktionalisierungen verwendet werden kann. Die so entstanden Hydroxyl α-Diazocarbonyl-Verbindungen wurden dann in metallkatalysierten, intramolekularen O–H-Insertionsreaktionen verwendet. Diese Reaktionen lieferten im Fall von $Rh_2(OAc)_4$ als verwendeter Katalysator zu Ausbeuten von bis zu 81%. Wurde hingegen $CuOTf \cdot Tol_{1/2}$ als Katalysator verwendet, resultierte die Reaktion in einer Diastereomerenselektivität von 7:93.

2   Introduction

The carbon-heteroatom (C–X) bond is a ubiquitous motif existing in natural and artificial molecules. The field of chemical synthesis has therefore attracted a lot of attention in recent years. The palladium-catalyzed C–N and C–O bond couplings, developed primarily by BUCHWALD and HARTWIG, provided an important method to construct this type of bonds.[1] Another class of carbon-heteroatom transformation that has undergone significant development in recent 10 years is the insertion of carbenoid into X–H bond. In these reactions, a metal-carbenoid is typically generated *in situ* when diazo precursor decomposed by a transition metal,[2] which can then undergo a variety of transformations, e.g. X–H (X = C, N, O, S, B, etc.) insertion, cyclopropanation, ylide formation, 1,2-migration as well as olefination and polymerization (Figure 1).[3–6]

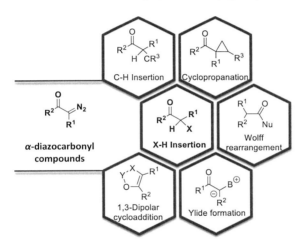

Figure 1. Small excerpt of the versatility of α-diazocarbonyl compounds. X = any heteroatom, but limited to N, O in this work; Nu = nucleophile; B = base; R^1 = alkyl, aryl, cycloalkyl. R^2 = alkyl, aryl, cycloalkyl, alkoxy, alkyl amino.

2.1   Synthesis of α-Diazocarbonyl Compounds

The first recorded synthesis of α-diazocarbonyl compound is ethyl diazoacetate (EDA) reported by CURTIS in 1883.[7] Although in 1912 WOLFF discovered the diazocarbonyl rearrangement which now bears his name,[8] simple diazocarbonyl compounds synthesis became readily available in the late 1920s. ARNDT and co-workers [9] as well as BRADLEY and co-workers[10] reported the acylation of diazomethane with an acid chloride to obtain these compounds. In recent years, various methodologies have been developed and there are now several well-established procedures to prepare different types of α-diazocarbonyl compounds (Figure 2).

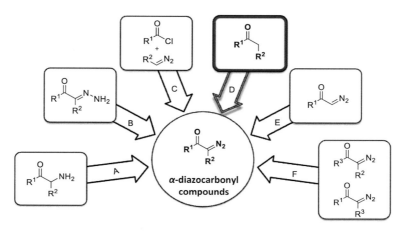

Figure 2. Common approaches to generate α-diazocarbonyl compounds: **A)** amine diazotization; **B)** modification of oximines; hydrazones and tosylhydrazones; **C)** acylation of diazoalkanes; **D)** diazo transfer; **E)** diazo cross-coupling; **F)** substituent modification.

Among them, acylation of diazoalkanes (**C**) as well as diazo transfer to an acid derivative or a ketone (**D**) are two basic methods to synthesize α-diazocarbonyl compounds and continue to evolve with new reagents and procedures. Metal-catalyzed cross-coupling at the diazo carbon (**E**) and substituent modification at either side (**F**) of the diazocarbonyl group can be added as approaches whereby a diazocarbonyl compound is converted into another one with retention of the diazo function.

2.1.1 Acylation of Diazoalkanes

Despite the hazardous nature of diazomethane,[11] acylation of diazoalkanes remains the most important route to the synthesis of terminal α-diazoketones. Acyl halides are usually the preferred acyl derivatives for diazoketone formation, but conversion of highly hindered carboxylic acids to the corresponding acyl chlorides can sometimes prove difficult or be impossible. Efforts have been made to activate the acids *via* acyl mesylates,[12] N-isocyanotriphenyliminophosphorane,[13,14] acyloxyphosphonium salts[15] (Scheme 1) and so on under milder reaction conditions, thus various functional groups could be tolerated.

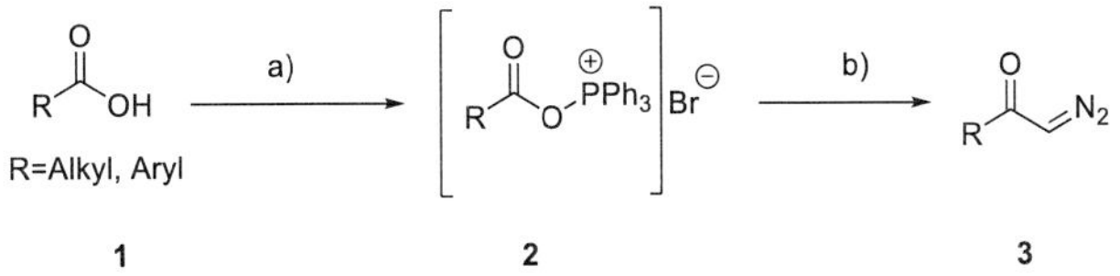

Scheme 1. Acylation of diazomethane *via* acyloxyphosphonium salt **2**: *a) PPh$_3$, NBS, THF, 0 °C, 15 min; b) CH$_2$N$_2$, r.t., overnight.*

2.1.2 Diazo Transfer Reactions

The diazo transfer technique was firstly introduced by REGITZ and his collaborators in 1967 to overcome the limitation of diazoalkane acylation,[16] which is not applicable to synthesize cyclic α-diazo ketones. Generally speaking, diazo transfer refers to the transfer of a diazo group from a donor to an acceptor, which for α-diazocarbonyl products must therefore be acids or ketone derivatives. The diazo donor are invariably sulfonyl azides.

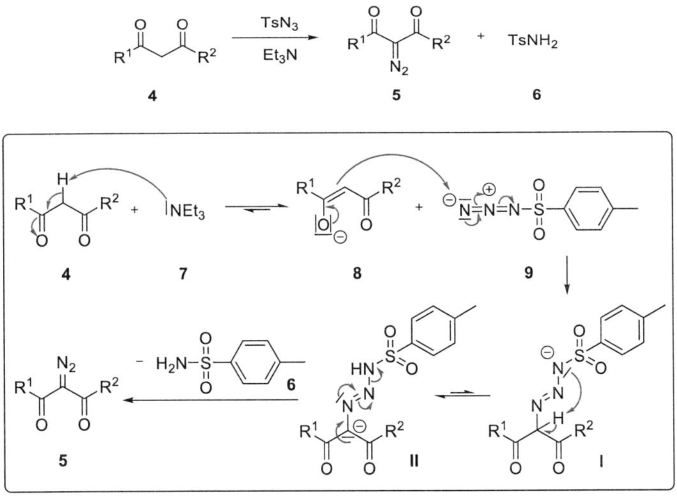

Scheme 2. Diazo transfer reaction of 1,3-dicarbonyl **4** mediated by tosyl azide (REGITZ Diazo Transfer) and its corresponding mechanism.

In order to transfer a diazo group to the α-methylene position of a carbonyl compound, the presence of a sufficient strong base to deprotonate the substrate is required. Thus, those with sufficiently reactive methylene groups toward diazo transfer, like malonic esters, β-keto esters and β-diketones can be readily converted to correspond 2-diazo-1,3-dicarbonyl products **5** under the standard REGITZ procedure by using triethylamine as base (Scheme 2).

The mechanism shows, by addition of triethylamine, the 1.3-dicarbonyl derivative **4** is first deprotonated at α-position to form the corresponding enolate **8**, which then attacks the diazo donor **9** due to its nucleophilic character. In transition state **I**, the nitrogen of the azide bound to the sulfur of tosylate and abstracts the second azide proton on the carbon. The negative charge **II** produced at the carbon atom then causes tosylamide **6** to be split off as a leaving group and remains the diazo group on the carbon to form the 2-diazo-1,3-dicarbonyl product **5**.

Simple ketones, without electron-withdrawing or aromatic group at the β-position, can be converted to the corresponding diazocarbonyl compounds in two steps involving a trifluoroacetyl activation followed by a diazo transfer reaction. The method was reported in 1990 by DANHEISER and is particularly interesting for the preparation α,β-unsaturated diazo ketones (Scheme 3, **A**).[17]

Taking advantage of DANHEISER's methodology, simple carboxylic esters can also be converted to the corresponding diazocarbonyl compounds, but the requirement of a strong base and cryogenic conditions may cause sometimes problems. To address these issues, TABER and co-workers developed the procedure of activating the ester *via* a titanium chloride-mediated benzoylation followed by diazo transfer under mild conditions (Scheme 3, **B**).[18]

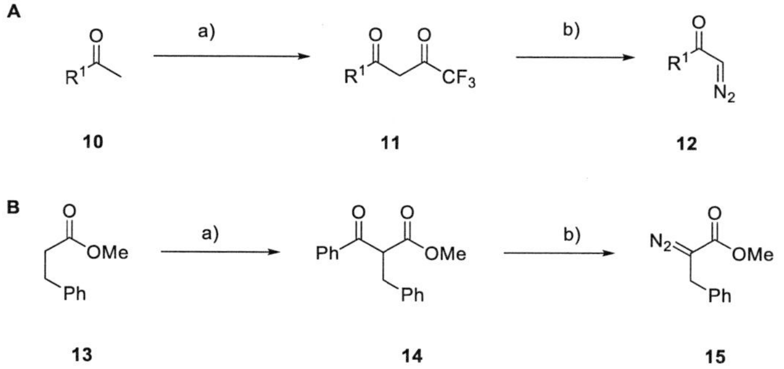

Scheme 3. Diazo transfer reaction *via* **A)** trifluoroacetylation/detrifluoroacetylation: *a) LiHMDS, CF₃CO₂CH₂CF₃, THF, −78 °C, b) R–SO₂N₃, base, CH₃CN, r.t.;* **B)** benzoylation/debenzoylation: a) *PhCOCl, TiCl₄, CH₃CN, 0 °C, b) R-SO₂N₃, base, CH₃CN, 0 °C.*

Besides deprotonation reagents, thermal stability of the transfer reagent and the nature of the byproduct are also two key parameters in the development of novel diazo transfer reagents. Since the byproducts tosylamide **6** can normally be removed by single filtration, the formal parameter is then more imperative (Figure 3).

The p-toluenesulfonyl azide **9** was first used to the synthesis of diazocyclopentadiene in 1953 by DOERING and co-workers and is still frequently employed as a diazo donor.[19] However, the chemists from MERCK[20] have raised doubts regarding the safety aspects of this reagent in large scale diazo transfer processes in their synthesis of THIENAMYCIN.[21] From the standpoint of utility, thermal stability, ease of handling as well as safety and cost, five reagents **16–20** have been subsequently examined. Among them, p-carboxyl-benzenesulfonyl azide (p-CBSA, **16**) has the highest initiation temperature, but the high price and the large loading of base made it not ideal for base sensitive substrates. p-Dodecylbenzenesulfonyl azide (p-DBSA, **17**), which was actually an isomeric mixture within the alkyl side chain, exhibited the lowest specific heat of decomposition and no impact sensitivity at the highest test level, making it the safest reagent of the group. DAVIES and his collaborators claimed that p-acetamidobenzenesulfonyl azide (p-ABSA, **18**) is a practical and cost-effective reagent in diazo-transfer reactions.[22] One of the more recent additions to the list is imidazolsulfonyl azide hydrochloride **21**,[23] which could be prepared in one-pot reaction on large scale from inexpensive materials and is shelf-stable. Further modifications by replacement of the hydrochloride counter ion in **21** with tetrafluoroborate in **22** or hydrogen sulfate in **23** have significantly improved the stability of these reagents in relation to shock sensitivity.[24]

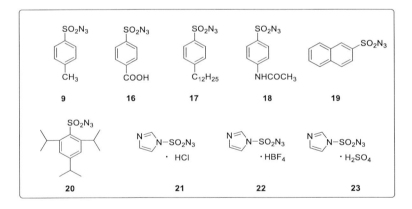

Figure 3. Attempts in designing the "ideal" sulfonyl azides.

2.1.3 Diazo Cross-Coupling Reaction

Due to the property that diazo carbon bears a partial negative charge, α-diazocarbonyl compounds can be also utilized as carbon nucleophiles in transition-metal-catalyzed coupling reactions. In 2007, WANG and co-workers reported the first synthesis of aryl or vinyl diazoacetates *via* palladium-catalyzed deacylative cross-coupling of aryl or vinyl iodides with ethyl diazoacetates at room temperature.[25] In 2015, the same group revealed the optimized reactions as following: 0.50 equiv. of silver carbonate as additive and triethylamine as base in toluene at room temperature (Scheme 4).[26] The present reaction conditions proved efficient over a wide range of substrates, which provided a practical method to introduce diazo functionality to organic compounds.

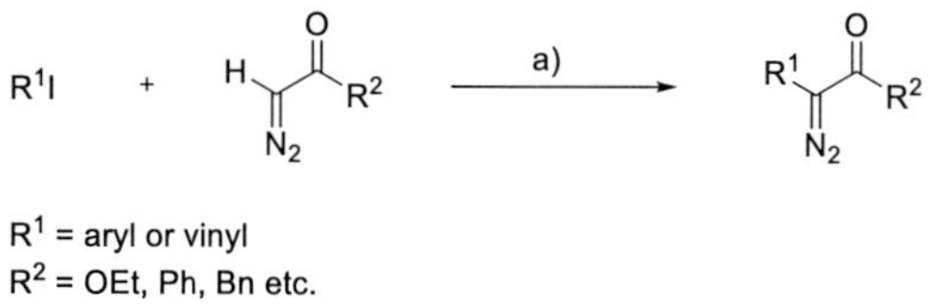

R^{1}I + H $\overset{O}{\underset{N_2}{\diagup\!\!\diagdown}}$ R^2 $\xrightarrow{\text{a)}}$ R^1 $\overset{O}{\underset{N_2}{\diagup\!\!\diagdown}}$ R^2

R^1 = aryl or vinyl
R^2 = OEt, Ph, Bn etc.

Scheme 4. Palladium-catalyzed cross-coupling of aryl or vinyl iodides with ethyl diazoacetate. Reaction conditions: *a) Pd(PPh₃)₄ (5 mol%), Ag₂CO₃ (0.5 equiv.), Et₃N, toluene, r.t., 4 h.*

2.2 Application of α-Diazocarbonyl Compounds in Metal-Catalyzed X–H (X = N, O) Insertion Reactions

Catalytic asymmetric insertion of carbenoids generated from α-diazocarbonyl compounds into X–H bonds have proven to be an efficient method to construct versatile and useful enantio-enriched building blocks. The generally accepted insertion mechanism includes the formation of the metal carbene intermediate *via* a transition-metal-mediated decomposition of diazo compound and the insertion of the electron-deficient metal carbene into a X–H (Scheme 5, **A**).[4] Two distinct pathways exist in the X–H insertion reactions. Depending on the polarity of the X–H bond, the mechanism can be divided into a concerted and a stepwise fashion. A concerted mechanism is generally accepted for reactions of carbenoids with non-polar bonds (*i.e.* cyclopropanation, C–H insertion or Si–H insertion).[27] However with polar X–H bonds, the weight of evidence prefers the stepwise ylide mechanism (Scheme 5, **B**).

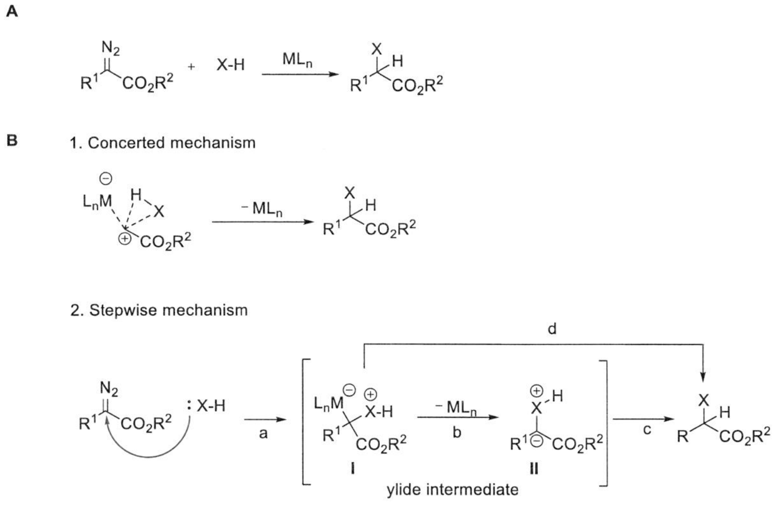

A

B   1. Concerted mechanism

2. Stepwise mechanism

Scheme 5. Transition-metal-catalyzed X–H insertion reactions and the corresponding mechanisms. **A**) Generally accepted mechanism of X–H insertion. **B**) Two distinct pathways exist in X–H insertions: 1. Reactions of carbenoids with non-polar bonds (X = C, Si). 2. Reaction of carbenoids with polar bonds (X = O, N, S).

The story of transition metal-catalyzed X–H insertion can be traced back to 1952, when YATES found α-hydroxyl substituted products instead of the intended WOLFF rearrangement product after α-diazoketones reacted with alcohols in the presence of copper salts (Scheme 6, **A**).[28] After the chemists from MERCK demonstrated in 1980 that an intramolecular N–H insertion can be used in large scale industrially to produce THIENAMYCIN (Scheme 6, **C**)[21] and the seminal observations from MOODY showed that O–H insertion can be widely used in accessing C–O bonds.[29] The X–H insertion reactions would have been thought to be quickly refined, but its development in the next two decades was sporadic. The major reason is that the polar X–H bonds could also be achieved through classical uncatalyzed nucleophilic substitutions. However, with the development of catalysis in the past few years, the prospect of chemo-, diastereo-, enantio-, regio-, and site-selective reaction became realizable, which is hard to achieve by classical uncatalyzed methods. Thus metal-catalyzed X–H insertion reactions got another opportunity.

2.2.1 Early Achievements in X–H Insertion Reactions

As mentioned above, YATES has employed finely developed copper catalysts and described the first systematic study of X–H insertion reactions.[28] He showed that a variety of X–H bond, like thiophenol, aniline, piperidine and ethanol can react with α-diazoketones to deliver corresponding products (Scheme 6, **A**). NOYORI and co-workers later came out with an explicit model for the copper-stabilized carbene in their groundbreaking papers on asymmetric catalysis.[30] Two seminal research were reported in 1973 in the metal-carbenoid catalysis filed. First, KOCHI and co-workers established that, Cu(I) is the catalytically active oxidation state in Cu-carbenoid chemistry.[31] Second, rhodium(II) acetate [Rh2(OAc)4] was successfully used as an excellent catalyst in the insertion of diazo compounds into hydroxylic bonds by TEYSSIE and co-workers (Scheme 6, **B**).[32] With the exceptional reactivity and high turnover of Rh2(OAc)4, this study initiated a new phase in metal-carbenoid chemistry , making it a subject of intense study for X–H insertions in the following few years.[2] Recently, YANG and co-workers exploited Rh2(OAc)4-catalyzed intramolecular O–H insertion in the total synthesis of Maoecrystal V[33] (Scheme 6, **D**), which again proved the X–H insertion can be utilized as a simple and elegant method to generate carbon-heteroatom bonds.

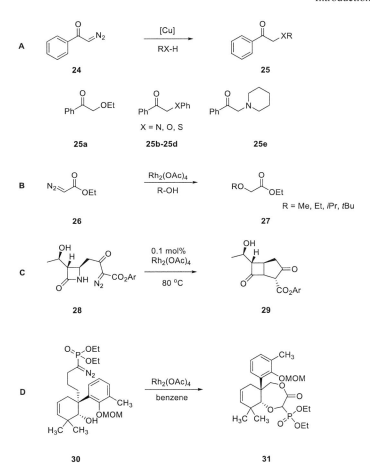

Scheme 6. The historical achievements in metal-catalyzed insertion of carbenoid into X–H bonds. **A**) Copper-catalyzed N–H, O–H and S–H insertion reactions. **B**) Rh₂(OAc)₄ as an excellent catalyst in O–H insertion reactions. **C**) Rh₂(OAc)₄-catalyzed intramolecular N–H insertion to produce THIENAMYCIN industrially. **D**) The example of Rh₂(OAc)₄-catalyzed intramolecular N–H insertion as a key step towards the total synthesis of Maoecrystal V.

2.2.2 Mechanistic Studies of Copper-Catalyzed X–H Insertion Reactions

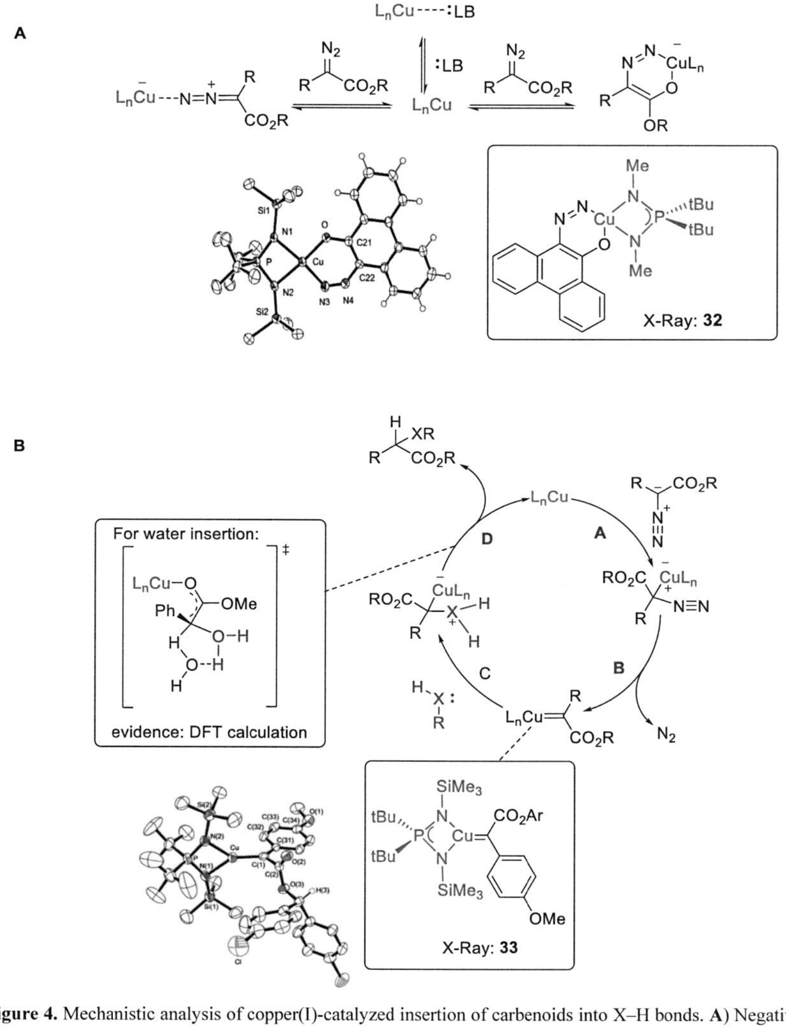

Figure 4. Mechanistic analysis of copper(I)-catalyzed insertion of carbenoids into X–H bonds. **A)** Negative metal carbene coordination modes. **B)** Commonly accepted mechanism for the metal-catalyzed X–H insertion reactions and evidence of step B and D as rate-determining step.

The commonly accepted mechanism for the X–H insertion reaction includes three main processes: metal-catalyzed nitrogen extrusion from α-diazocarbonyl compound to form metal-carbenoid (Figure 4, **B**, step A and B), generation of metal-associated oxonium ylide from metal-carbenoid and X–H bond (Figure 4, **B**, step C) and [1,2]-hydrogen shift of the metal-associated ylide to give the C–X bond and regenerate the metal catalyst (Figure 4, **B**, step D).

According to density functional theory (DFT) calculations, copper-catalyzed X–H insertion reaction has similar transition state energies in step A, B and D.[34,35] Therefore the rate-determining step (RDS) of copper (I) catalyzed O–H insertion reactions varies depending on the specific copper salt and ligand set employed. This is indeed also corroborated by the experimental results (Figure 4). Taking a closer look at step A and B to generate the metal-carbenoid, we can clearly find some negative factors, like potential coordination modes of substrates or LEWIS bases can slow down or even halt copper catalysis by blocking the crucial substrate binding site. This has been proven by the coordination mode from crystal structure (Figure 4, **32**) obtained with α-diazophenanthrenone.[36] Coordination of the negatively polarized carbon of the diazo substrate to the LEWIS acidic free copper (Figure 4, **B**, step A) is the first productive step in the catalytic cycle. The release of nitrogen to deliver the copper carbenoids have been characterized spectroscopically (Figure 4, **33**) from HOFMANN and co-workers.[37] When being treated with olefins, many of these stable carbenoids undergo rapid cyclopropanation, which thus confirmed the relevance of copper carbenoids in the catalytic cycle.

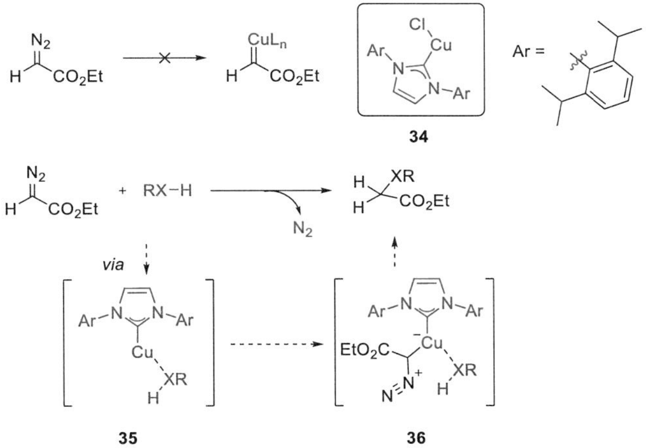

Scheme 7. The experimental evidence of the step D in the commonly accepted mechanism of metal-catalyzed X–H insertion reaction (Figure 4): the substrate-containing complex as active species.

FRUCTOS and co-workers have observed that the copper(I)-N-heterocyclic carbene complex **34** has no activity towards ethyl diazoacetate (EDA) without the participation of the other substrate containing X–H bond.[38] Only when both substrates are present, the reaction shows a rapid consumption of EDA ensue, leading to the expected X–H insertion or cyclopropanation products (Scheme 7). An explanation for these observations might be that coordination of the substrate lowers the energy of the transition state **35** leading to copper carbenoid **36**. These observations confirmed the importance of step D (Figure 4, **B**) in the catalytic cycle as well as highlighted the ability of ligands to dictate the course of catalytic reactions.

2.2.3 Mechanistic Studies of Rhodium (II)-catalyzed X–H Insertion Reactions[39]

Due to the difficulty to detect the intermediates in the rhodium (II)-catalyzed insertion reactions, most of the current mechanic understanding comes from computational studies and kinetic analysis. PIRRUNG and co-workers found that only one of the two available axial coordination sites is catalytically active at any given time through a detailed kinetic experiment of various rhodium(II) complexes (Scheme 8).[27]

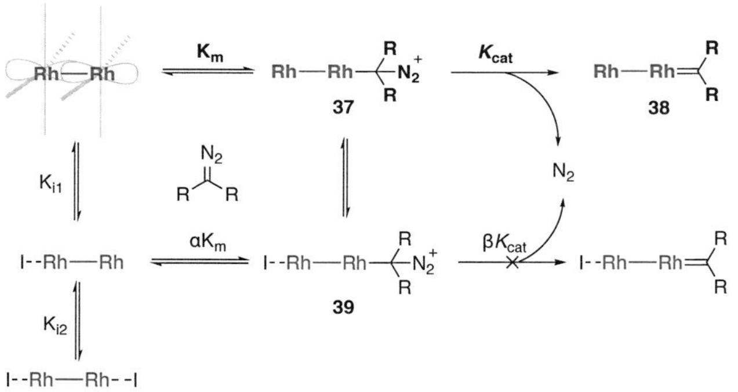

Scheme 8. Kinetic analysis of rhodium-catalyzed X–H insertion reaction and the effects of inhibitors. I = inhibitor.

Although their research is not specific for X–H insertion reactions, the results are pertinent. In copper-catalyzed X–H insertion reactions, the LEWIS base can slow down or even halt copper catalysts by blocking the crucial substrate binding site (Figure 4, **A**), which can be therefore considered as the potential inhibitory functional group here.

Briefly, the starting rhodium (II) complex has two vacant axial coordination sites, which can either be inhibited by LEWIS base (K_{i1} and K_{i2}) or bind to substrate (K_m). The substrate-bound rhodium complex (**37**, path K_m) can then lose nitrogen to deliver rhodium carbenoid (**38**, path K_{cat}). Alternatively, when one axial site is bound by an inhibitor and the other by a substrate molecular, the result rhodium-complex could also ultimately lead to the requisite rhodium carbenoid (**39**, path αK_m), but this pathway does not contribute to catalysis (βK_{cat}). Moreover, the pathway αK_{i1} is a minor contributor to the total reaction flux as it works just at inhibitor concentrations of 50 mM or even less. Hence, most catalytic turnovers occur according to the $K_m \rightarrow K_{cat}$ pathway. It is also important to note that loss of nitrogen (path K_{cat}, Scheme 8) is the RDS in all reactions they examined. Although the ylides in rhodium(II)-catalyzed X–H insertion reactions have never been reported, their relevance can be inferred from examples where the putative ylide intermediates have been trapped with imine electrophiles[40] or even coaxed into [2,3]-sigmatropic rearrangement pathways[41] rather than the [1,2]-proton shift. These results provided not only evidence for the existence of ylide intermediates, but also suggest in line with computational results,[34] that the [1,2]-proton shift has a surprisingly high activation barrier.

2.2.4 Ruthenium-catalyzed X–H Insertion Reactions

Ruthenium, one of rhodium's direct neighbors in the periodic table, has an electron configuration of $4d^75s^1$ and the widest scope of oxidation states ($-$II in $Ru(CO)_4^{2-}$ to octavalent in RuO_4), as well as various coordination geometries in each electron configuration. However, as a consequence of the difficulties in matching catalysts and substrates, the application of ruthenium chemistry has been limited to oxidations with RuO_4, hydrogenation reaction and hydrogen transfer reactions until the 1980s.[42]

The first ruthenium porphyrin complex catalyzed ethyl diazoacetate insertion into S–H and N–H bonds was reported in 1997.[43] Recently, CHE and co-workers reported the one-pot carbenoid N–H insertion using $[RuCl_2(p\text{-cymene})]_2$ as catalyst.[44] LACOUR and co-workers have shown that a combination of $[CpRu(CH_3CN)_3][PF_6]$ **40** and 1,10-phenanthroline **41**, can efficiently catalyze O–H (Scheme 9) insertion and condensation reactions of α-diazocarbonyl compound **42**.[45]

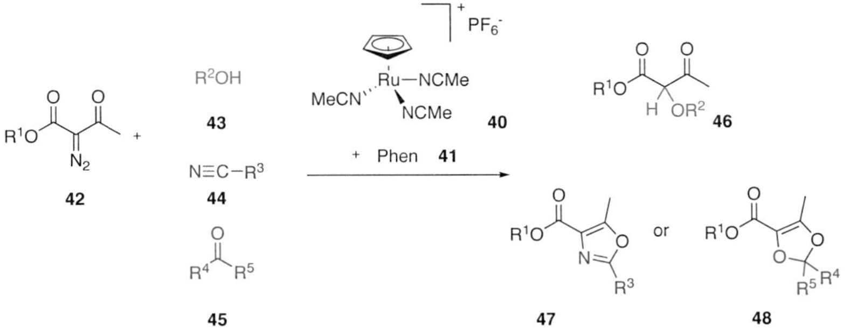

Scheme 9. Ruthenium-catalyzed O–H insertion and condensation reactions of α-diazocarbonyl compound **42** in the presence of $[CpRu(CH_3CN)_3][PF_6]$ and 1,10-phenanthroline. Cp = cyclopentadienyl.

These results demonstrate that ruthenium complexes can sometimes offer unique reactivity in comparison with cooper and rhodium. Although as a relative newcomer in the metal-catalyzed insertion reactions, the favorable properties of ruthenium, like similar reactivity as rhodium, lower cost, more available oxidation states, and rich coordination chemistry would suggest a bright future of their application in this field.

2.3 Development of Enantioselective X–H Insertion Reactions

2.3.1 Enantioselective N–H Insertion Reactions

The earliest example of asymmetric N–H insertion was reported by KAGAN through a chiral auxiliary approach in 1971.[46] With substrates containing chiral ester side-chains, the copper cyanide-promoted N–H insertion led to amino acid products in up to 26% ee. Due to the discovery of the outstanding catalytic abilities of rhodium (II) acetate [Rh₂(OAc)₄], research in this field has been mainly focused on the development of efficient chiral rhodium catalysts. The research on copper-catalyzed X–H insertion reactions was therefore idled for almost 20 years. Meanwhile, MCKERVEY and co-workers reported the first asymmetric intramolecular N–H insertion in the presence of chiral rhodium(II) carboxylates in 1996, which afforded pipecolic acid derivatives with 45% ee.[47] With chiral dirhodium(II) catalysts, MOODY and co-workers reported an intermolecular N–H insertion between methyl α-diazophenylacetate or dimethyl α-diazobenzylphosphonate and benzyl carbamate, which gave products in good yields but only 9% ee was accessed.[48] In 2004, JØRGENSEN and co-workers obtained up to 28% ee by using copper(I)-bisoxazoline complex **49** (Scheme 10), which led chemists re-evaluated the comprehensive ability of copper catalyst.[49]

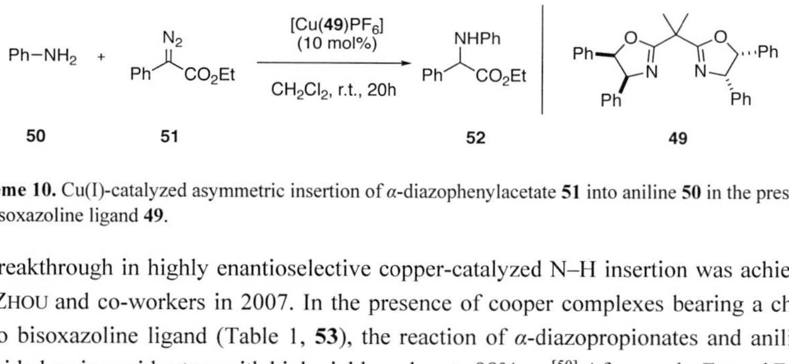

Scheme 10. Cu(I)-catalyzed asymmetric insertion of α-diazophenylacetate **51** into aniline **50** in the presence of bisoxazoline ligand **49**.

A breakthrough in highly enantioselective copper-catalyzed N–H insertion was achieved by ZHOU and co-workers in 2007. In the presence of cooper complexes bearing a chiral spiro bisoxazoline ligand (Table 1, **53**), the reaction of α-diazopropionates and anilines provided amino acid esters with high yields and up to 98% ee.[50] Afterwards, FU and FENG reported two other copper-catalyzed asymmetric N–H insertion reactions with bispyridine (Table 1, **54**)[51] and chiral binaphthyl derivatives (Table 1, **55**)[52] as ligands, respectively. Both catalysts performed good yields and excellent enantioselectivities.

Table 1. Copper-catalyzed asymmetric N–H insertion from α-diazocarbonyl compounds.

53	54	55
x = 5, y = 6	x = 7, y = 8	x = y = 10
R^1 = Me	R^1 = aryl	R^1 = Me
R^3 = aryl	R^3 = Boc or Cbz	R^3 = aryl
86-96% yield	48-89% yield	90-99% yield
85-98% ee	80-95% ee	87-98% ee

Although rhodium (II) complexes have dominated the field of catalytic X–H insertion reactions for nearly two decades, they suffered from the poor ability in promoting the enantioselectivity of X–H insertion reactions compared to copper(I) complexes. The DFT calculation of O–H insertion carried out by YU'S group showed that whereas copper complexes prefer to remain bound to the substrate during protonation (Figure 4, **B**, step D), the rhodium catalysts favor dissociation and hence cannot transfer chiral information.[34] Despite the existence of such a "weakness", enantioselective N–H insertion based on rhodium(II) catalyst have been successfully realized in 2011. ZHOU and co-workers reported the asymmetric N–H insertion of α-diazocarbonyls with *tert*-butyl carbamate (*t*-BocNH$_2$)[53] as well as benzyloxycarbonyl carbamate (CbzNH$_2$)[54] cooperatively catalyzed by dirhodium(II) carboxylates and chiral spiro phosphoric acids (SPA). Under optimized conditions, up to 99% yield and 95% ee was obtained (Scheme 11). The proposed mechanism demonstrated that the achiral rhodium(II) carbene can be intercepted by an imine activated intermediate with a chiral BRØNSTED acid **56**, which indicates that cooperative rhodium(II)-BRØNSTED acid catalysis may offer an alternative approach towards enantioselective X–H insertions.

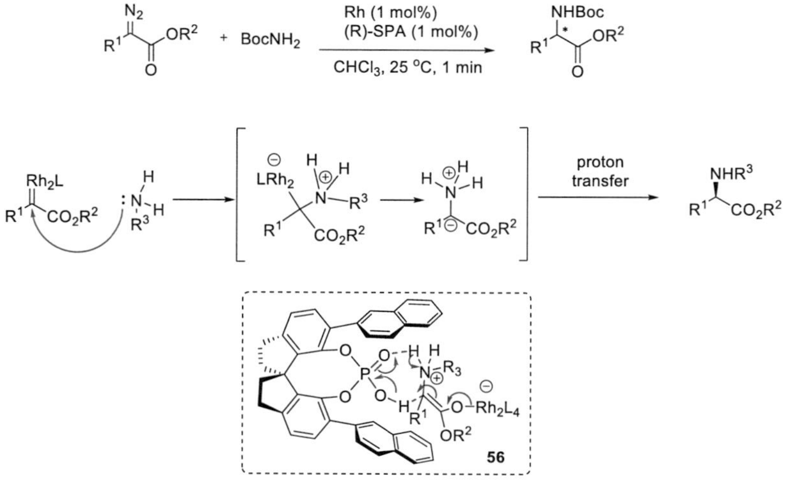

Scheme 11. Rhodium-catalyzed enantioselective N–H insertion of α-diazocarbonyl compounds with carbamates: proposed mechanism *via* a cooperative rhodium(II)-BRØNSTED acid catalyst approach.

2.3.2 Enantioselective O–H Insertion Reactions

The history of O–H insertion reactions of α-diazocarbonyl substrates parallels closely to their N–H insertion counterparts. The most significant accomplishments in enantioselective O–H insertion have been made in the last decade. Earlier efforts by MOODY and co-workers to devise enantioselective or diastereoselective processes having met with only limited success.[55] A breakthrough of cooper-bisazaferrocene-catalyzed highly enantioselective O–H insertion was achieved by FU and co-workers in 2006 (Table 2, **57**), which generated α-alkoxyesters in high yields and up to 98% ee.[56] A more simple and direct approach to the enantioselective synthesis of chiral α-aryl-α-hydroxy/alkoxyesters was subsequently uncovered by ZHOU and co-workers with the discovery of the well-defined copper-[50] and iron-[57]spiro bisoxazoline catalysts (Table 2, **53**). The catalyst structure studies from ZHOU suggested that, the phenyl groups on the oxazoline ligands could form a perfect C_2-symmetric chiral pocket around the copper center to induce the chirality.[58]

Based on this conception, MUKAI and co-workers reported the first C_2-symmetric planar-chiral [2.2]paracyclophane-based bisoxazoline ligands (Table 2, **58**) in 2015, which characterized by the inserted benzene as spacer to offer not only the conformational flexibility, but also the steric or electronic element based on the benzene to interact with substrate and/or the reactant.[59] The catalyst generated up to 80% ee in copper-catalyzed O–H insertion of α-diazocarbonyls without the aid of the central chirality.

Table 2. Copper-catalyzed enantioselective O–H insertion reactions from α-diazocarbonyl compounds.

57	53	58
x = 2.0, y = 3.8	x = 5.0, y = 6.0	x = 5.0, y = 6.0
R^1 = alkyl, aryl	R^1 = aryl	R^1 = Et, Ph
R^2 = aryl	R^2 = alkyl, aryl	R^2 = Me, Ph
56-98% yield	62-88% yield	53-92% yield
up to 98% ee	up to 99% ee	up to 80% ee

In conclusion, the highly enantioselective insertion of α-diazocarbonyl compounds into X–H bond was first achieved with copper catalyst. Due to the outstanding reactivity and high turnover of rhodium acetate performed in N–H insertion reactions, rhodium (II) catalysts were exclusively employed until the end of the nineties. Ruthenium (II) and iron (III) catalysts have been frequently studied in recent years and shown unique reactivities. Given the long history, it is not hard to be found that the metal salt and the ligand set employed in the reaction plays a crucial role in the reactivity and enantioselectivity. Thus, the design and synthesis of efficient and universal chiral ligands, preparation of stable metal complexes as well as searching for appropriate reaction conditions are still the challenge for the future research.

2.4 Paracyclophane Derivatives in Asymmetric Catalysis

2.4.1 [2.2]Paracyclophane

[2.2]Paracyclophane (**59**) (PC),[60] which was first named as 'di-*p*-xylylene', was prepared and isolated by BROWN and FARTHING in 1949. The systematic investigation of its unique chemical behavior was carried out by CRAM in 1951.[61] The structure displays an abnormally small distance between two phenyl rings (para position: 2.78 Å, ring center: 3.10 Å) compared to normally stacked aromatics (3.40 Å) (Figure 5).

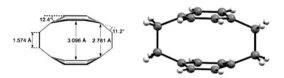

Figure 5. Structural features of [2.2]paracyclophane.

The two parallel aromatic rings offer an interesting platform to investigate π–π interactions and constitute bricks for the construction in polymer and material science.[62] Another interesting feature is the chirality of easily available mono-substituted PCs and their application in asymmetric catalysis.[63] Moreover, planar-chiral derivatives of [2.2]paracyclophane undergo racemization only at a relatively high temperature, and their cyclophane backbone is chemically stable towards light, oxygen, acids and bases.[64]

2.4.2 Chirality of [2.2]Paracyclophane Derivatives

The presence of one substituent breaks the initial D_{2h} symmetry of [2.2]paracyclophane and as a consequence brings chirality.[65] The assignment of absolute configuration (Figure 6) and numbering can be achieved by taking advantage of CAHN-INGOLD-PRELOG (CIP) rules. One has to determine the 'pilot' atom as the nearest atom out of the chiral plane, the latter is the plane bearing the atom with the highest priority according to CIP-rules. A detailed description is shown in chapter 6.1

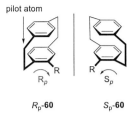

R_p-**60** S_p-**60**

Figure 6. Absolute configuration determination of mono-substituted paracyclophane.

Considering the disubstituted paracyclophane derivatives, particular descriptors are used to designate the relation between the two positions. For substituents on the same phenyl ring, *ortho*, *meta* and *para* prefixes are still valid, for inter-ring disubstitution pseudo-*gem*, pseudo-*ortho*, pseudo-*meta* and pseudo-*para* are employed as prefixes. The disubstituted paracyclophane with two identical substituents can be also planar-chiral according to specific substitution pattern (Figure 7).[66–70]

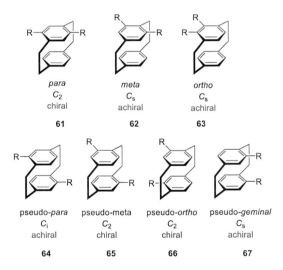

Figure 7. Homo-disubstituted [2.2]paracyclophanes with different substitution patterns.

2.4.3 Functionalization of [2.2]Paracyclophane Derivatives

The mono-substituted [2.2]paracyclophane derivatives can be obtained *via* classic electrophilic aromatic substitution (S_EAr) reactions. Some well-established methods, like bromination,[71–76] nitration,[77] FRIEDEL-CRAFTS acylation[78,79] as well as formylation[80,81] are frequently used as the initial step. A variety of functional groups, like amines, carbonyl acids, hydroxyls and further modification can be obtained from these compounds.

The disubstituted PCs are considerably more complicated compared to the mono substitutions. Among the four possible disubstitution patterns, pseudo-*gem* disubstitution is represented in terms of its specific transannular directive effect (Scheme 12).[82] Theoretically, the repulsion between the two proximate substituents makes the structure sterically unfavored, but the LEWIS basic group (E), which could be introduced after the above-mentioned S_EAr reaction, is ideally placed to capture the pseudo-*gem* proton. The WHELAND intermediate **69** is generated *via* an intramolecular deprotonation process, which makes the pseudo-*gem* disubstitution kinetically favorable.

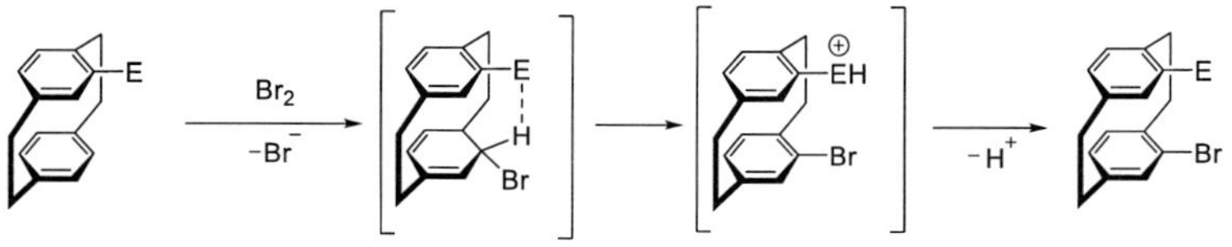

E = COOH, CO_2R, C(=O)R, C(=O)NR_2, NO_2, SO_2R, POR_2, oxazoline

Scheme 12. Synthesis of pseudo-*gem* disubstituted [2.2]paracyclophane derivatives: a kinetically favored WHELAND intermediate obtained *via* the transannular directive effect.

The main source of pseudo-*meta* disubstituted PC-derivatives is the dibromoparacyclophane, which is obtained through neutral bromination conditions (Scheme 13) from unsubstituted [2.2]paracyclophane. Based on different solubilities, the major product **72** can be isolated in 43% yield after recrystallization in chloroform. The remaining solid **71** can be obtained in 34% yield after recrystallization in dioxane, which provides a method to obtain pseudo-*para* dibromo [2.2]paracyclophane.[83] Pseudo-*gem* isomer **74** can also be transformed to the pseudo-*meta* isomer under a thermal isomerization protocol (Scheme 13). The inherent mutual repulsion of the two substituents on pseudo-*gem* compounds drives the equilibrium almost totally towards the thermodynamically favored pseudo-*meta* isomer **72**.[82]

Pseudo-*ortho* [2.2]paracyclophane derivatives are more convenient to be prepared from the pseudo-*para* isomer by thermal isomerization or microwave irradiation (Scheme 13). The protocol developed by BRADDOCK using microwave-assisted isomerization provided a fast and efficient method to transform pseudo-*para* dibromo-PC **71** into pseudo-*ortho*-dibromo-PC **75** after several runs.[84]

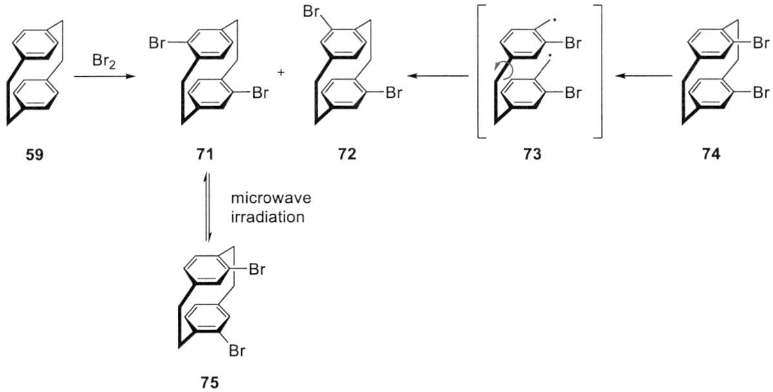

Scheme 13. Synthesis of *pseudo*-dibromo[2.2]paracyclophanes *via* different routes.

Another type of functionalization is the coordination of [2.2]paracyclophane as π-base with transition metals. Because of its enormous abundance of electrons, [2.2]paracyclophanes are better π-electron donors than the corresponding phenyl derivatives, and the positive charge on the transannular arranged second phenyl ring can thus be better stabilized. These properties have been confirmed by CRAM and WILKINSON during their investigations of tricarbonylchromium complexes with different [m.n]paracyclophanes (Figure 8, **76**).[85] These complexes are formed not only faster compared to benzene derivatives, but also only monocomplexes are observed when m.n ≤ 4. The monochromium tricarbonyl complexes of substituted [2.2]paracyclophanes have been reported later, which also showed the preferred coordination of chromium carbonyl to the electron-rich ring as well as the evidence of an inter electronic interaction between the two rings.[86,87] Several years later ELSCHENBROICH and co-workers successfully synthesized bis(η^6-[2.2] paracyclophane) chromium(0) complex **77** and (η^{12}-[2.2] paracyclophane) chromium (0) **78** by condensation of [2.2]paracyclophane with chromium atoms.[88]

In the 1980s, BOEKELHEIDE and co-workers have reported various ruthenium(II) complexes of [2.2]paracyclophane as η^6-arene ligands. Besides monomeric structures **79**, bis(η^6-[2.2]para-cyclophane) ruthenium(II) complexes **80** and dinuclear ruthenium complexes **81** can also be obtained.[89–94]

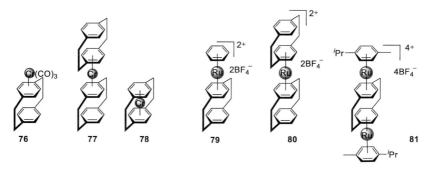

Figure 8. Chromium(0) and ruthenium(II) complexes from [2.2]paracyclophane as η⁶-arene ligands.

2.4.4 Application of [2.2]Paracyclophane Derivatives in Asymmetric Catalysis

The application of [2.2]paracyclophane in asymmetric catalysis was first reported by PYE and ROSSEN in 1997. The PHANEPHOS (**82**)-rhodium (I) catalyzed hydrogenation from unsaturated amino acid derivatives **83** provided product **84** in quantitative yields as well as excellent enantioselectivity under very mild condition (Scheme 14).[95] PHANEPHOS (**82**)-ruthenium catalyst has been successfully used in enantioselective hydrogenation of β-ketoesters **85**. The β-hydroxyesters **86** were obtained in quantitative yields and up to 96% ee under 0.8 mol% catalyst loading[96].

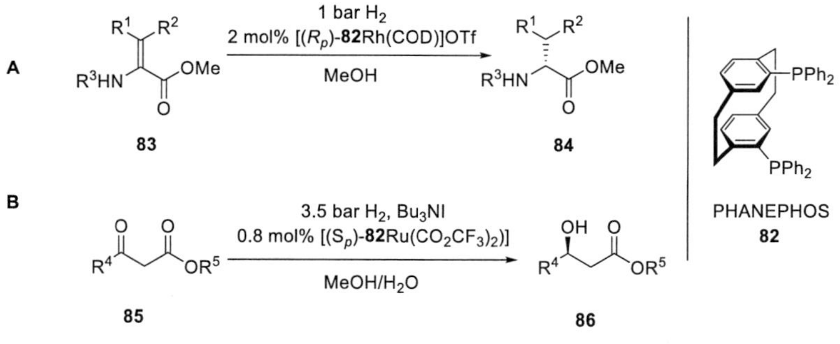

Scheme 14. Application of PHANEPHOS **82** in asymmetric hydrogenation. **A)** PHANEPHOS-rhodium(I)-catalyzed enantioselective hydrogenation. **B)** PHANEPHOS-ruthenium(II)-catalyzed hydrogenation from β-ketoesters **85**.

With the advantages of bench stability, excellent activity and high enantioselectivity, PHANEPHOS has been widely used in industrial-scale hydrogenations.[97] Based on the success of PHANPHOS-rhodium-catalyzed enantioselective hydrogenation, a number of derivatives as well as complexes coordinated with different metals have been studied. PHANEPHOS-dipalladium complexes showed up to 91% ee in the syntheses of chiral carboxylic esters, which provided an exciting method for hydrocarbonylation and hydroesterification.[98] The analogues of PHANEPHOS, xyl-PHANEPHOS-platinum catalyst was found to be favorable in asymmetric oxidative cation/alkyne cyclization to access polyenes.[99]

P,N-bidentate ligands are often more stable than *P,P*-ligands and are amenable to the introduction of additional stereochemistry through the nitrogen moiety. Among them, chiral oxazoline-substituted planar-chiral [2.2]paracyclophanes have attracted great attention from chemists. The first phosphinooxazoline-based [2.2]paracyclophane ligand **87** was reported by REID in 2000 based on the study of pseudo-*gem* directed bromination.[100] This type of *P,N*-ligands were later successfully applied in palladium-catalyzed enantioselective allylic alkylation. Under optimized condition, the (R_p,*R*)-**87** provided the target product **90** in 98% yield and 90% ee (Scheme 15).[101,102]

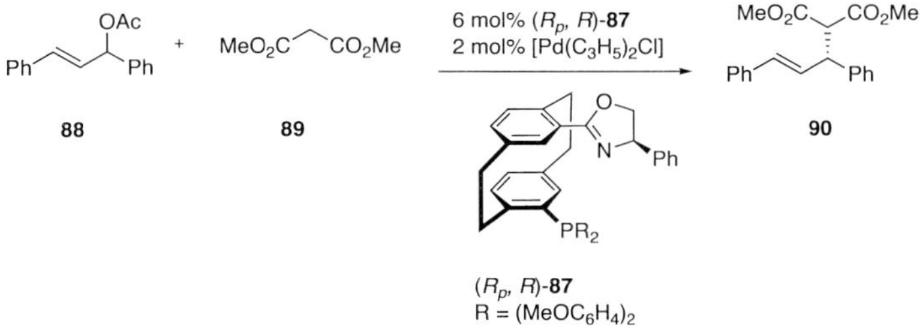

Scheme 15. *P,N*-[2.2]paracyclophane-palladium-catalyzed asymmetric allylic alkylation.

Due to the easy preparation from readily available chiral amino alcohols and high reliability for asymmetric induction,[103] chiral bisoxazoline (box) ligands have been applied to diverse metal-catalyzed asymmetric reactions. Box ligands containing only the central chirality, some other ligands developed by combining the central and axial chirality (biaryls and naphthyls,[104,105] chiral spiro[4,106,107]) or central and planar ones (biferrocene[108,109]) have been studied. Among them, the chiral spiro bisoxazoline ligands have been widely used in metal-catalyzed X–H insertion reactions[4,53,54,57,58,110,111] and performed excellent reactivity and enantioselectivity. [2.2]paracyclophane, as a typical example of planar chiral ligands, has been successfully used in various asymmetric catalysis and showed great configurational stability under harsh conditions. However, the [2.2]paracyclophane-based bisoxazoline ligands have not been reported until 2015. The ligands synthesized by MUKAI *et al.* are characterized by the inserted benzene as spacer, which provided the conformational rigidity and flexibility to make the distance between the two functional groups suitable for performing. The steric or electronic elements (Scheme 16, R on **58**) based on the benzene ring could interact with the substrates and/or the reactants.[59] The ligands showed up to 80% ee in copper-catalyzed O–H insertion from α-diazoesters (Scheme 16).

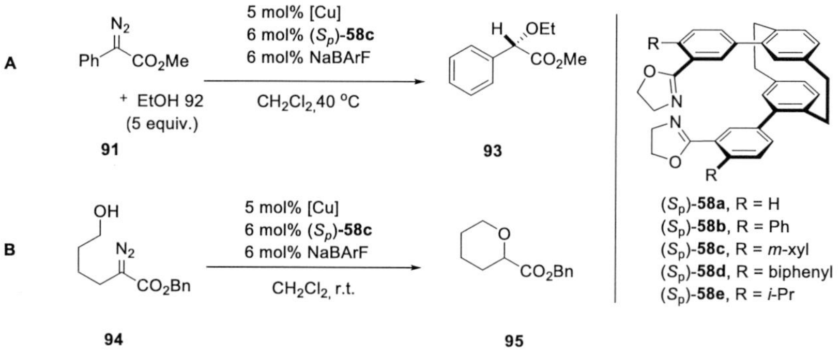

Scheme 16. The applications of planar chiral [2.2]paracyclophane-based bisoxazoline ligands in metal-catalyzed X–H insertion reactions. **A)** Cu-catalyzed intermolecular aliphatic O–H insertion from methyl 2-diazo-2-phenylacetate **91** in the presence of (S_p)-**58c**. **B)** Cu-catalyzed intramolecular aliphatic O–H insertion from **94** in the presence of (S_p)-**58c**.

Compared to the disubstituted [2.2]paracyclophane derivatives mentioned above, monosubstituted [2.2]paracyclophanes have received relatively less attention, due to their flexibility and not suitable for the introduction of chirality. Nevertheless, GALTZHOFER and co-workers have reported the monosubstituted [2.2]paracyclophane ligand **96** in copper-catalyzed enantioselective cyclopropanation of aromatic olefins by means of adding a bulky group around the copper center to create chiral pocket. Up to 90% conversion and 65% ee have been achieved for selected olefins (Scheme 17).[112]

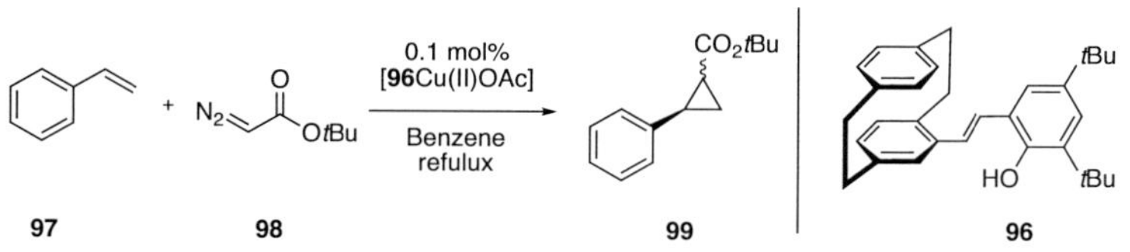

Scheme 17. Copper-catalyzed enantioselective cyclopropanation of aromatic olefins with *tert*-butyl 2-diazoacetate **98** in the presence of monosubstituted [2.2]paracyclophane ligand **96**.

In addition, HOU and co-workers reported the chiral [2.2]paracyclophane monophosphine ligands in palladium-catalyzed umpolung allylation of aldehyde with cyclohexenyl acetate, giving homoallyl alcohols in good to very good yields and up to 60% ee.[113] Although these ee values are exceeded by other chiral ligands, it provided an important information, that simple monosubstituted [2.2]paracyclophane can be also used in asymmetric catalysis.

2.5 Enzymes as Efficient Biocatalysts in the Asymmetric Reduction of Ketones

2.5.1 General Remarks and Potential of Enzymes as Catalysts in Organic Synthesis

Nature, the best chemist during the life evolution, has evolved a certain set of synthetic strategies and used an impressive array of enzyme catalysts to construct substances from simple metabolites to complicated natural products.[114] Nonetheless, enzyme was not realized at the beginning as a first choice in synthetic chemistry due to the existing disadvantages, like narrow substrate range, limited stability of enzymes under organic reaction conditions, low efficiency when using wild-type strains, diluted substrate and product solutions. However, due to the tremendous progress in enzyme discovery, enzyme engineering and process development in recent years, these disadvantages have been properly avoided. Enzymatic catalysis is increasingly recognized by organic chemists in both academia and industry as an attractive synthetic tool besides the "classic" synthesis, metal catalysis as well as organocatalysis.[115]

Like all proteins, enzymes are made up of a long chain of amino acids that is assembled and then folded into a complex three-dimensional structure (Figure 9).[116] Due to the structural characteristic and the active site integrated therein, the enzyme-catalyzed reactions have highly specific recognition of specific substrates, leading to excellent chemo-, regio-, diastereo- as well as enantioselectivity. These unique properties of enzyme to stereoselectivity recognize a substrate was first proposed by FISCHER in his *Lock-and-Key Hypothesis*,[117] according to which the substrate has to fit into the active site of the enzyme like a key into the lock. In addition to the most value of high enantio- and diastereoselectivity, enzymes are considered "renewable catalysts" as they are produced by the fermentation of bacteria using sugars and biodegradable once they have been used in a process. What is more, enzymatic catalysis is highly welcomed in industries, in particular the chemical and pharmaceutical industries, in terms of its green chemistry feature,[118] which the enzymes are generally of low toxicity, the enzymatic reactions are often performed in water at or slightly over room temperature and enzymes are renewable as mentioned above.

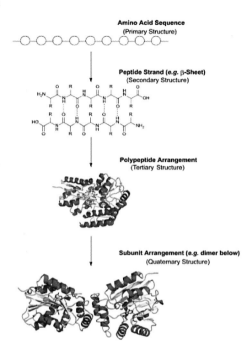

Figure 9. The structure of enzymes. Copyright© 2018 Royal Society of Chemistry.

2.5.2 Type of Enzymatic Reactions Commonly Used in Organic Synthesis

Though many different types of enzymes are used in nature to catalyze various of different transformations, only limited number of reaction types can be widely used in biocatalysis. According to the way they work on a molecular level, six major classes of enzymes in organic synthesis are summarized in the Table 3:

Table 3. Summary of the major classes of enzyme and their common use.

Enzyme class	Name	Common uses
EC1	Oxidoreductse $A_{red} + B_{ox} \rightarrow A_{ox} + B_{red}$	catalyze oxidation or reduction reactions, which involve the transfer of electrons from the reductant to the oxidant
EC2	Transferases $A{-}B + C \rightarrow A + B{-}C$	Catalyze the transfer of a functional group from one molecule to another
EC3	Hydrolases $A{-}B + H_2O \rightarrow A{-}H + B{-}OH$	Breaking the chemical bonds with addition of water (hydrolysis)
EC4	Lyases $A(XH){-}B \rightarrow A{-}X + B{-}H$	Generating a double bond
EC5	Isomerases $A \Leftrightarrow И_{30}{-}A$	Catalyze structural changes within a molecule
EC6	Ligases $A + B + ATP \rightarrow A{-}B + ADP + Pi$	Join to two molecules

Each enzyme is assigned with two names. The first is its short, recommended name, which has the suffix "-ase" attached to the substrates of the reaction, for example, glucosidase, urease; or to a description of the action performed, like lactate dehydrogenase and ketoreductase. The second is the systematic name, which is much more complicated, but used when the enzyme must be identified without ambiguity. Each enzyme has a code number (EC) that characterizes the reaction type (first digit), subclass (second digit), and subsubclass (third digit). The forth digit is for the specific enzyme.[119]

2.5.3 Enzyme-Catalyzed Asymmetric Reduction of Ketones

Ketones are widespread in nature and are often connected with other functional groups. The carbonyl group can be interconverted into a range of other functional groups and the oxygen atom of carbonyl group is capable of forming hydrogen bond interactions with other functional groups. The asymmetric reduction of ketones has the potential to produce chiral alcohols, which are valuable intermediates and building blocks in active pharmaceutical ingredients (APIs), agrochemicals and natural products.[120] There are several traditional synthetic methods for the asymmetric reduction of ketones, such as COREY-BAKSHI-SHIBATA (CBS) reduction,[121–123] which converts the linear or cyclic ketones into corresponding alcohols with highly enantioselectivity by using the chiral oxazaborolidine reagent. However, the synthesis of ligand is required, and care needs to be taken not to produce hydrogen gas in large scale reactions. Another alternative is the use of the metal-catalyzed hydrogenation with a chiral ligand.[124] The reactions are highly enantioselective, but the cost of metal, ligand synthesis as well as the metal waste disposal need to be considered.

Compare to chemical methods, the biocatalytic ketone reduction possesses unparalleled chemo-, regio- and stereoselectivity. Moreover, the enzyme is easily disposed at the end of the reaction without any specialized treatment. Ketoreductases (KREDs) or alcohol dehydrogenase (ADHs), classified under EC 1.1.1, are co-factor-dependent enzymes, which use a nicotinamide adenine dinucleotide as co-factor. The co-factor existed in two forms: the phosphorylated version (NADPH), whilst the non-phosphorylated version is NADH. The nicotinamide co-factor is the hydride source, which the enzyme uses to carry out the reduction of the substrate. The byproduct of the reduction is NAD^+ or $NADP^+$ depending on whether it is phosphorylated or not.[116] In the course of enzymatic reduction with the majority of ADHs, the nicotinamide moiety of the co-enzyme approaches the substrate molecular in an orientation in which the amide group is situated in front of the less-bulky substitute of the substrate. The carbonyl group of the substrate is then attacked by the hydride from the top (*re*face), which is called *pro-R* hydride (H_R in Figure 10) to give the corresponding alcohol in (*S*)-configuration. This process is described by PRELOG's rule. When the ketone is attacked by the hydride from *si*face, (*R*)-alcohol is consequently produced, which is named as *anti*-PRELOG product. A few ADHs are known to have anti-Prelog stereopreference, *Lactobacillus kefir* (LkADH) is one of which.[125]

In the asymmetric reduction of a prochiral ketone, there are four possible hydride transfer pathways to deliver the hydride from NAD(P)H, as shown in Figure 10. The ADH-catalyzed transfer of *pro-S* hydride (H_S) to the *re* side of the carbonyl is not known yet.[126] Hydrogen sources are necessary to perform the reaction. Since the coenzyme NAD^+ or its phosphate $NADP^+$ are costly due to their high hydride equivalent weight, a successful regeneration of enzyme is crucial to make the reaction practical. A cheap and readily available cosubstrate, such as 2-propanol or ethanol can be used in the reaction system, which not only plays the role as hydride source, but also increases the solubility of organic substrates.

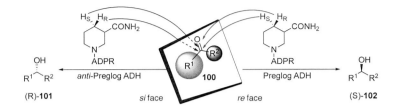

Figure 10. Four possible hydride transfer pathways from NAD(P)H to the carbonyl carbon on a ketone substrate. R^1 is more sterically hindered and has higher CAHN-INGOLD-PRELOG priority than R^2, ADPR = adenosine diphosphoribose.

3 Aim of the Work

Transition-metal-catalyzed X–H insertions (X = C, N, O, S, B and Si) *via* metal carbenes or carbenoids to generate carbon-heteroatom (C–X) bonds have received tremendous attention in the past decade.[4,5,25,26,50,53,54,56–58,111,127] Although recent progress on copper-catalyzed enantioselective insertion of carbenoids into N–H bonds in the presence of spiro bisoxazoline ligand **53** has met a great success, many challenges continue to limit the application of this important reaction. For instance, this reaction is highly substrate-dependent: good yields and excellent enantioselectivities are obtained from simple α-diazo substrates. Compare to spiro bisoxazoline ligands, the [2.2]paracyclophanes-based bisoxazoline ligands bearing central and planar chirality, should perform great configurational stability and enantioselectivity under similar reaction system. The here presented work focuses on the synthesis of diverse complicated α-diazocarbonyl compounds and their application in the enantioselective N–H insertion reactions with [2.2]paracyclophane-based bisoxazoline ligands (Figure 11, **A**). In addition to metal-catalyzed insertion reactions, the enzymatic synthesis of highly enantioselective alcohols followed by O–H insertion to form oxygen-containing heterocycles is also targeted (Figure 11, **B**).

In the first part of the project, various unsaturated α-diazocarbonyl compounds **103** should be synthesized as precursors for metal-catalyzed N–H insertion and intramolecular O–H insertion reactions to generate amino acid derivatives and *O*-containing heterocycles, respectively .

Furthermore [2.2]paracyclophane-based bisoxazoline ligands, namely (S_p)- or (R_p)-pseudo-*ortho* [2.2]paracyclophane-based bisoxazolines should be accessed and applied in metal-catalyzed N–H insertions in order to obtain good reactivity and enantioselectivity. The obtained δ-amino α,β-unsaturated carboxylic esters **104** are then supposed to undergo further functionalizations to build diene-dienophile precursors for intramolecular DIELS-ALDER reactions to generate hexahydroindoles, which are key intermediates towards the total synthesis of Rostratin B–D.

The last part of the project is a cooperation with Prof. CHRISTOF NIEMEYER'S group, from the Institute for Biological Interfaces (IBG-1) in the KIT to realize the chemoenzymatic synthesis of *O*-containing heterocycles. The aim was to enantioselectivity reduce the keto-contained α-diazocarbonyl compounds **109** to obtain chiral alcohols with the help of ketoreductases. The optimal intramolecular O–H insertion condition starting from racemic alcohols should be found and then implemented to a consecutive reaction sequence in a flow reactor to enable large scale synthesis of the desired product **110** with good diastereo- and enantioselectivity (Figure 11, **B**).

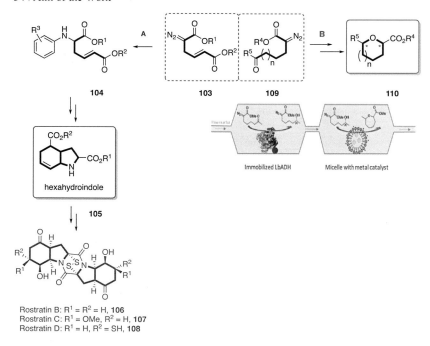

Figure 11. Aim of the project: **A)** Synthesis and application of α-diazocarbonyl compounds in metal-catalyzed N–H insertion reactions in the presence of [2.2]paracyclophane-based bisoxazoline ligands to generate δ-amino α,β-unsaturated carboxylic esters **104** and further functionalization to access hydroindoles **105**. **B)** Enzymatic reduction of keto-contained α-diazocarbonyl compounds followed by O–H insertion to generate O-containing heterocycles.

4 Results and Discussion

The focus of this work lies on the synthesis and application of α-diazocarbonyl compounds. The results are divided into three parts, including the synthesis of α-diazocarbonyl compounds, the metal-catalyzed N–H insertion in the presence of [2.2]paracyclophane-based bisoxazoline ligands and the enzymatic approach to obtain enantiopure α-diazocarbonyl compounds as well as their application in metal-catalyzed intramolecular O–H insertion to obtain O-containing heterocycles.

4.1 Synthesis of α-Diazocarbonyl Compounds

4.1.1 α-Diazocarbonyl Compounds for Model System Tests

To investigate the metal-catalyzed N–H insertion reactions in the presence of [2.2]paracyclophane-based bisoxazoline ligands, a series of simple α-diazocarbonyl compounds that known in the literature were chosen as starting materials for model system tests. Starting from substituted acetoacetates (R^1 = Et, tBu, Bn), the corresponding products were obtained *via* nucleophilic substitution[128] followed by a classic diazo transfer reaction reported by REGITZ *et al.*[16] with up to 95% yield (Scheme 18). The low yields of **113a** and **113b** may due to the volatility of simple α-diazocarbonyl compounds, although both were purified *via* flash column chromatography using low boiling point solvents pentane and Et$_2$O as eluent.

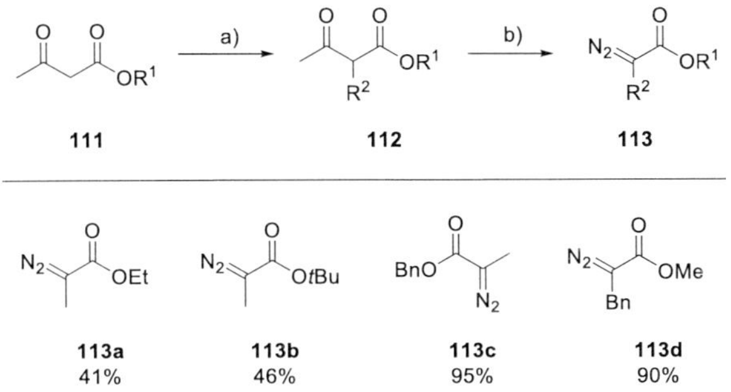

Scheme 18. Synthesis of α-diazocarbonyl compounds **113a–113d** *via* diazo transfer method. Reaction conditions: *a) RI, NaH, abs. THF, reflux, overnight; b) p-ABSA, Et$_3$N, CH$_3$CN, 0 °C to r.t.*

Phenyl 2-diazopropanoate **116** was synthesized *via* amine diazotization since phenyl 3-oxobutanoate (R^1 = Ph) is not commercial available and the synthesis needs to undergo high temperature with 63% reported yield after purification.[129] Under amine diazotization condition, the precursor **115** can be synthesized *via* a simple nucleophilic substitution[128] with 97% yield and the product **116** was obtained with 21% yield as yellow oil after purification (Scheme 19).

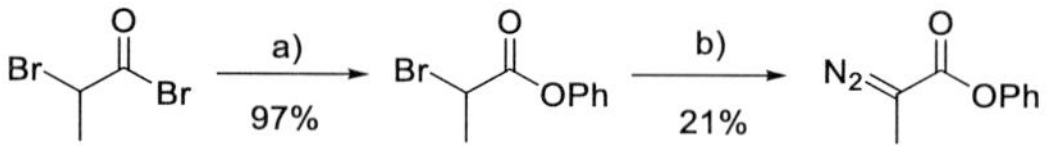

114 **115** **116**

Scheme 19. Synthesis of phenyl 2-diazopropanoate **116** *via* amine diazotization. Reaction conditions: *a) phenol, pyridine, 0 °C, 30 min, 97%; b) TsNHHNTs, DUB, THF, 0 °C, 21%.*

By taking advantage of amine diazotization, methyl 2-diazo-2-phenylacetate **118** can be synthesized directly from methyl 2-oxo-2-phenylacetate **117** in 79% yield (Scheme 20).

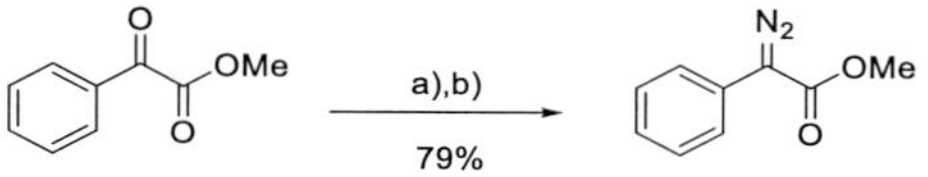

117 **118**

Scheme 20. Synthesis of methyl 2-diazo-2-phenylacetate **118** *via* amine diazotization. Reaction conditions: *a) $TsNHNH_2$, toluene, reflux, 9 h; b) Et_3N, CH_2Cl_2, r.t., 36 h.*

4.1.2 α-Diazocarbonyl Compounds for Enzyme-Catalyzed Reduction

In order to obtain the chiral alcohol to perform chemoenzymatic synthesis of O-containing heterocycles, the keto-contained α-diazocarbonyl compounds **109** should be firstly accessed. Compared to the work mentioned above to generate methyl- (**112a–113c, 116**) and benzyl- (**113d**) substituted α-diazocarbonyl compounds, an additional step to synthesize different iodo-substituted ketones is necessary here (Scheme 21). The 5-iodopentan-2-one **120a**, 6-iodohexan-2-one **120b** and 4-iodo-1-phenylbutan-1-one **120c** were obtained *via* FINKELSTEIN reaction in 94%, 66% and 93% yield, respectively. The products can be used directly after extraction and drying under vacuum since the only byproduct sodium chloride is not soluble in acetone and can be therefore removed by single filtration.

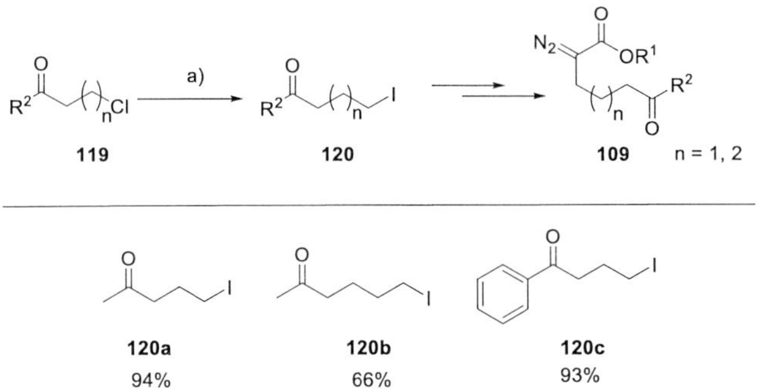

Scheme 21. Synthesis of iodo-substituted ketones **120** *via* FINKELSTEIN reaction. Reaction conditions: a) *NaI, acetone, reflux, overnight.*[130]

With the iodo-substituted ketones in hand, six different substituted acetoacetates (Table 4, entry 1–6, **121a–121f**) were obtained *via* nucleophilic substitution in 21–48% isolated yields. Consequently, six related α-diazocarbonyl compounds (Table 4, entry 1–6, **109a–109f**) can be obtained *via* diazo transfer reaction in moderate yields (entry 1–5). The low yields for **121f** and **109f** probably arise from the relative bulky phenyl group, compared to the methyl substituted derivatives.

Table 4. Summary of the synthesized keto-contained α-diazocarbonyl compounds **109** for the consequent chemoenzymatic synthesis of O-containing heterocycles.

Entry	n	R¹	R²	Yield 121 [%][a]	Yield 109 [%][a]
1	1	Me	Me	48 (**121a**)	60 (**109a**)
2	1	Et	Me	34 (**121b**)	46 (**109b**)
3	1	Bn	Me	34 (**121c**)	59 (**109c**)
4	2	Bn	Me	37 (**121d**)	54 (**109d**)
5	1	Me	Ph	38 (**121e**)	44 (**109e**)
6	1	Bn	Ph	21 (**121f**)	34 (**109f**)

Reaction conditions: *a) NaH, THF, reflux, overnight; b) p-ABSA, Et₃N, 0 °C to r.t., overnight.* [a] Isolated yields after column chromatography.

4.1.3 Unsaturated α-Diazocarbonyl Compounds for the Synthesis of Hexahydroindole: Key Intermediate towards the Total Synthesis of Rostratin B-D

The C_2-symmetric thiodiketopiperazine natural products Rostratin B–D (**106–108**) consist of two identical octahydroindole amino acid moieties and are bridged with a disulfur unit. The synthetic challenge they present and their potent activities against viruses, bacteria, and cancer cells[131–133] make them particularly interested in total synthesis.[12,134–137]

According to the retrosynthesis route (Scheme 22), the disulfur unit should be introduced into the diketopiperazine core **122** in the last stage. The latter should be obtained *via* a microwave-assisted dimerization from a free octahydroindole amino acid **123**. The free octahydroindole amino acid monomer **123** can be accessed *via* functionalization from the hexahydroindole precursor **105**, which can be generated *via* an intramolecular DIELS-ALDER reaction from the diene-dienophile system **124**.[138–140] The diene-dienophile **124** can be generated *via* a sequence of oxidative deprotection, condensation with crotonaldehyde and protection of the δ-amino α,β-unsaturated carboxylic ester **104**. In a previous work, BRÄSE *et al.* reported a three-component vinylogous MUKAIYAMA-

MANNICH reaction of 2-furfural **125**, *p*-anisidine **165c** and ethyl (*E*)-but-2-enoate **127** with optimized solvent and temperature to generate unsaturated amino acid precursors in large-scale preparations (Scheme 22).[136]

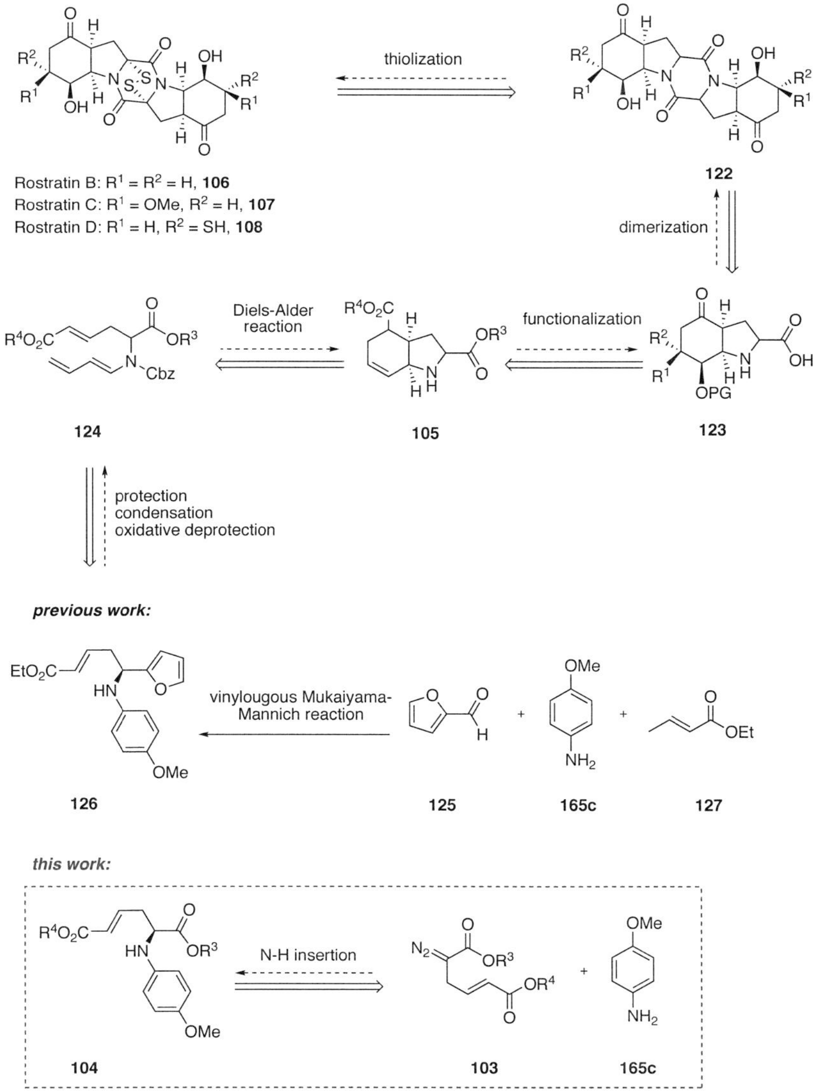

Rostratin B: R^1 = R^2 = H, **106**
Rostratin C: R^1 = OMe, R^2 = H, **107**
Rostratin D: R^1 = H, R^2 = SH, **108**

124

105

123

protection
condensation
oxidative deprotection

previous work:

126

125

165c

127

this work:

104

103

165c

Scheme 22. Retrosynthesis approach of Rostratin B–D. PG = protecting group. Cbz = carboxybenzyl.

Based on the importance of hexahydroindole **105**, in which the unsaturated double bond provides a platform for further functionalization to access the amino acid monomer **123** in the total synthesis, a metal-catalyzed N–H insertion from unsaturated α-diazocarbonyl compounds with aniline is carried out in this work to build the carbon skeleton. The detailed description will be shown in chapter 4.3.3.

Therefore, the first challenge in front of us was the synthesis of unsaturated α-diazocarbonyl compounds. Based on the successful experience we obtained from the synthesis of saturated α-diazocarbonyls, the same condition was applied with 4-bromocrotonate **128** and methyl acetoacetate **129** in the presence of sodium hydride as base (Scheme 23).

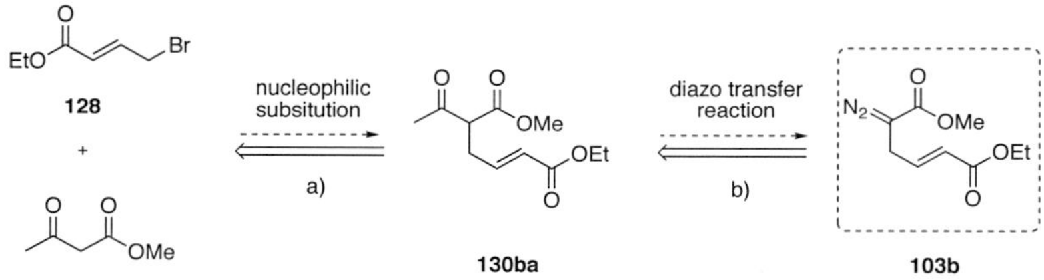

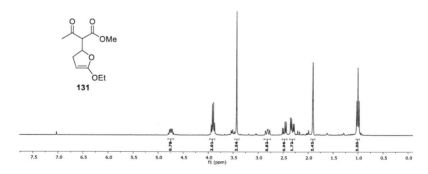

Scheme 23. Designed approach to generate **103b** *via* nucleophilic substitution followed by diazo transfer reaction. Reaction condition: a) *NaH, THF, 0 °C to reflux, 2 h; b) p-ABSA, Et₃N, CH₃CN, r.t., overnight.*

Figure 12. ¹H NMR spectrum of the proposed MICHAEL addition byproduct **131**. – **¹H NMR** (300 MHz, CDCl₃) δ/ppm = 5.06 – 4.88 (m, 1H, C*H*CH₂), 4.14 (q, *J* = 7.1 Hz, 2H, C*H*₂CH₃), 3.66 (s, 3H, OC*H*₃), 3.19 – 2.92 (m, 1H, C*H*COCH₃), 2.71 (dd, *J* = 15.9, 7.4 Hz, 1H, C*H*=), 2.54 (dd, *J* = 15.7, 6.1 Hz, 2H, CHC*H*₂), 2.14 (s, 3H, COC*H*₃), 1.23 (t, *J* = 7.2 Hz, 3H, CH₂C*H*₃).

Although a fraction with identical molecular weight (m/z = 228) as target product **130** was found from gas chromatography-mass spectrometry (GC-MS), the ¹H NMR showed no characteristic peak of a *trans* alkene[1]. According to the ^{1}H NMR spectrum (Figure 12), the multiple peak between 3.19 and 2.92 suggests the proton from a substituted methylene group, and the multiple peak from 5.06 to 4.88 might be a proton from a substituted alkene. Considering the α,β-unsaturated carbonyl substrate **128** as well as the basic reaction condition, the obtained byproduct **131** is supposed to be a MICHAEL addition product.

Inspired from the work reported by WANG and co-workers, who coupled ethyl diazoacetate (EDA) with vinyl iodides in the presence of palladium(0) catalyst,[25,26] we changed the starting material from ethyl (Z)-3-bromoacrylate **132** to ethyl 4-bromocrotonate **128** and carried out the reaction under the same condition (Scheme 24). However, no desired product was found after purification. We suppose the unsatisfied result is due to the limited scope of the methodology, which just vinyl iodides/bromides bearing electron-withdrawing groups work under this system.

- Previous work: WANG et al.[25,26]

- This work:

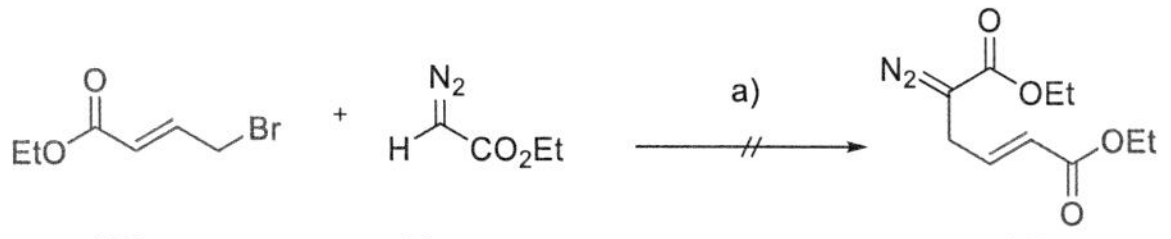

Scheme 24. Palladium-catalyzed cross-coupling from EDA. Reaction conditions a) $Pd(PPh_3)_4$ (5 mol%), Et_3N (1.50 equiv.), nBu_4NBr (1.00 equiv.), acetone, 35 °C, 2 h.

After these unsuccessful attempts, we reconsidered using the classic diazo transfer method as depicted in chapter 4.1.2 to generate the target compounds **103**. Taking into account the electron-deficient α,β-double bond of allenes bearing an electron-withdrawing group (EWG), while β,γ-double bond is electron-rich,[141] a phosphine-catalyzed γ-Umpolung addition of nucleophiles to 2,3-butanedienoates (Scheme 25) can be applied to synthesize the precursor **130**.[142–144]

---

[1] δ/ppm = 6.66 (dt, 1H,$^3J_{trans}$ = 15.6, ^3J = 7.1 Hz, CH$_2$C*H*=CH), 5.69 (dt, $^3J_{trans}$ = 15.6, ^4J = 1.5 Hz, 1H, CH$_2$CH=C*H*).

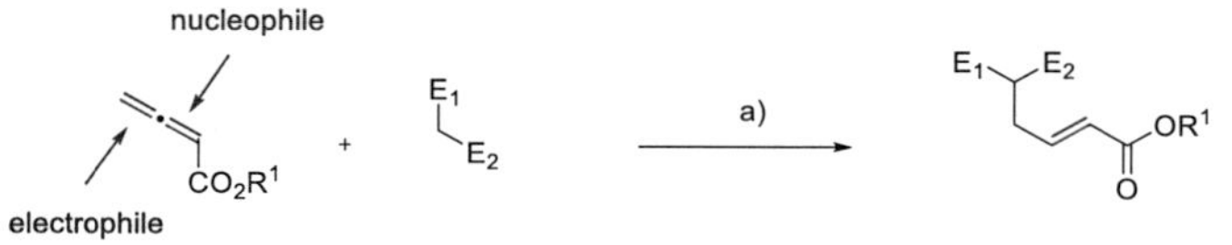

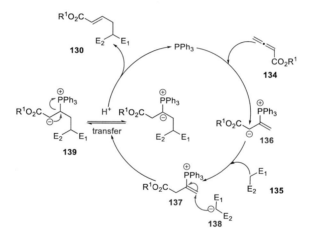

Scheme 25. Phosphine-catalyzed Umpolung addition reaction of nucleophiles **135** with 2,3-butandienoates **134**. Reaction condition: *a) PPh₃ (5 mol%), HOAc (0.50 equiv.), NaOAc (1.00 equiv.), benzene, 80 °C, 2h, 56%, E/Z > 97/3.*

A plausible mechanism is outlined in Figure 13.[142] Triphenylphosphine attacks the *β*-carbon of 2,3-butandienoates **134** to give the phosphonium intermediate **136**, which then deprotonates the pronucleophile **135** to form the corresponding vinylphosphonium **137** and the carbanion **138**. MICHAEL addition of the anion **138** to the vinylphosphonium salt **137** followed by a proton transfer to afford the adduct **130**.

Figure 13. Mechanism of the *γ*-Umpolung addition of nucleophiles **135** to 2,3-butanedienoates **134** catalyzed by triphenylphosphine.

The synthesis of **103** is outlined in Scheme 26. Starting from substituted 2-bromoacetate **141**, the corresponding triphenylphosphorane analogues (**142a–142d**) are obtained as off-white solids after nucleophilic addition and elimination in nearly quantitative yields according to an established protocol from BÄCKVALL *et al.*[145] The elimination step was carried out in a separation funnel, in which the obtained product could directly transfer to the organic phase as soon as the reaction finished in the aqueous phase.

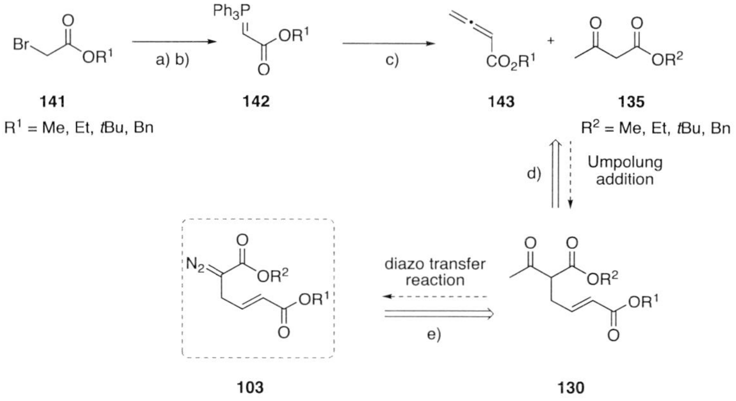

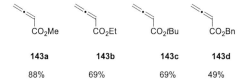

Scheme 26. Schematic synthesis route to generate the unsaturated α-diazocarbonyl compounds **103**. Reaction conditions: *a) PPh₃, EtOAc, r.t., overnight; b) aq. NaOH, CH₂Cl₂; c) AcCl, Et₃N, CH₂Cl₂/pentane, 0 °C to r.t., 6h; d) **135**, PPh₃, benzene, 80 °C, overnight; e) p-ABSA, KF, CH₃CN, 0 °C to r.t., overnight.*

After vigorously shaking, the resulting product was collected in the dichloromethane phase and used directly after drying under vacuum without further purification. The 2,3-butanedienoates (**143a–143d**) were synthesized *via* a WITTIG reaction using triethylamine as base and dichloromethane-pentane as co-solvent.[146–149] The byproduct triphenylphosphine oxide was formed as yellow-orange precipitate in pentane and removed by filtration. Due to the fact that methyl buta-2,3-dienoate (**143a**) and ethyl buta-2,3-dienoate (**143b**) are volatile, the purifications are carried out using low boiling point solvents as eluent [II] and the water bath of the rotary evaporator was set to 20 °C during removal of the eluent. A summary of synthesized buta-2,3-dienoates (**143a–143d**) are shown in Figure 14.

CO₂Me	CO₂Et	CO₂*t*Bu	CO₂Bn
143a	**143b**	**143c**	**143d**
88%	69%	69%	49%

Figure 14. Summary of the synthesized buta-2,3-dienoates **143a–143d**.

In the next step, the 2,3-butanedienoates (**143a–143d**) are treated with a variety of β-acetoacetates (**135a–135c**) in the presence of a catalytic amount of triphenylphosphine

---

[II] Column chromatography condition: pentane:ether = 3:1 (v/v).

in benzene under reflux (Table 5). The reaction of ethyl buta-2,3-dienoate **143b** and methyl acetoacetate **135a** proceeded smoothly to produce the γ-Umpolung product ($\rightarrow$**130ba**) in 68% yield as a single geometry (E)-isomer (Table 5, entry 2). Instead of a ratio of $E/Z = 97/3$ in the mixture reported in literature,[142] 5% of the tautomer from methyl acetoacetate **135a** was found together with **130ba** through the integration of a single peak at $\delta = 11.5$ ppm on the ^{1}H NMR spectrum. Good to fair yields were obtained from the reactions of **143a** with other β-acetoacetates (Table 5, entry 1, 3, 5). The reactions of methyl acetoacetate **135a** with *tert*-butyl buta-2,3-dienoate **143c** (Table 5, entry 4) and benzyl buta-2,3-dienoate **143d** (Table 5, entry 6) showed relatively low yields, the main reason could attribute to the steric hindrance around the dienoates.

Table 5. A summary of the synthesized α,β-unsaturated carboxylic ester substituted acetoacetate **130**.[a]

R^1 = Me, Et, *t*Bu, Bn
R^2 = Me, Et, *t*Bu, Bn

Entry	R^1	R^2	Yield [%][b]
1	Me (**143a**)	Me (**135a**)	43 (**130aa**)
2	Et (**143b**)	Me (**135a**)	68 (**130ba**)
3	Me (**143a**)	*t*Bu (**135b**)	42 (**130ab**)
4	*t*Bu (**143c**)	Me (**135a**)	33 (**130ca**)
5	Me (**143a**)	Bn (**135c**)	57 (**130ac**)
6	Bn (**143d**)	Me (**135a**)	38 (**130da**)

[a] Reaction conditions: 2,3-butandienoate **143a**–**143d** (1.00 equiv.), β-acetoacetate **135a**–**135c** (1.10 equiv.), PPh₃ (5 mol%), HOAc (0.50 equiv.), NaOAc (1.00 equiv.), benzene, 80 °C, 2 h. [b] Isolated yields after column chromatography.

An initial attempt to transfer diazo group to **130** using triethylamine as base and *p*-ABSA as diazo transfer reagent in acetonitrile resulted in only 12% yield (Table 6, entry 1). The yield could be slightly improved by changing the base to DBU (Table 6, entry 2). To our delight, with using potassium fluoride as base, the reaction afforded 74% yield (Table 6, entry 3). After determining the optimized base, other reaction parameters were subsequently screened. The reaction seemed to be favored in polar solvents such as acetonitrile compared to chloroform (Table 6, entry 3–4), but ethanol resulted in a diminished yield (Table 6, entry 5). The different diazo transfer reagents did not show a significant difference in yields (Table 6, entry 6–8), but due to the stability and ease preparation in the laboratory, *p*-ABSA was favored in the other tests. The successful use of potassium fluoride in this reaction is apparently due to the facile deprotonation of *β*-acetoacetates by the 'naked' anion of fluoride. Similar as potassium fluoride, tetra-*n*-butylammonium fluoride (TBAF) is widely used as a mild base to provide the source of fluorine. The good solubility of TBAF in organic solvents makes it a useful alternative to poorly soluble inorganic bases. Thus, TBAF was also tested in this reaction (Table 6, entry 9), however no product was found after stirring overnight under room temperature. K_2CO_3 yielded just trace amount of the product (Table 6, entry 10).

Table 6. Screening of conditions of the diazo transfer reaction to synthesize **103b**.[a]

Entry	Base	Azide	Solvent	Time	Yield [%][b]
1	Et₃N	*p*-ABSA	CH₃CN	overnight	12
2	DUB	*p*-ABSA	CH₃CN	overnight	20
3	**KF**	***p*-ABSA**	**CH₃CN**	**overnight**	**74**
4	KF	TsN₃	CHCl₃	overnight	12
5	KF	TsN₃	EtOH	5	42
6	KF	TsN₃	CH₃CN	5	59
7	KF	*p*-DBSA	CH₃CN	5	50
8	KF	*p*-ABSA	CH₃CN	5	54
9	TBAF	*p*-ABSA	CH₃CN	overnight	-[c]
10	K₂CO₃	*p*-ABSA	CH₃CN	overnight	4

[a] Reaction condition: **130ba** (1.00 equiv.), azide (2.00 equiv.), base (3.00 equiv.). [b] Isolated yields after column chromatography. [c] = no product formed.

Table 7. Summary of the synthesized unsaturated α-diazocarbonyl compounds **103a–103f**.

Entry[a]	130	103	Yield [%][b]
1	**130aa** (OMe, CO$_2$Me)	**130a** (N$_2$, OMe, CO$_2$Me)	56
2	**130ba** (OMe, CO$_2$Et)	**130b** (N$_2$, OMe, CO$_2$Et)	56
3	**130ab** (OMe, CO$_2$tBu)	**103c** (N$_2$, OMe, CO$_2$tBu)	60
4	**130ca** (OtBu, CO$_2$Me)	**103d** (N$_2$, OtBu, CO$_2$Me)	40
5	**130ac** (OMe, CO$_2$Bn)	**103e** (N$_2$, OMe, CO$_2$Bn)	68
6	**130da** (OBn, CO$_2$Me)	**103f** (N$_2$, OBn, CO$_2$Me)	37

[a] Reaction conditions: β-acetoacetate precursors **130** (1.00 equiv.), *p*-ABSA (2.00 equiv.), KF (3.00 equiv.) in CH$_3$CN, r.t., overnight. [b] Isolated yield after column chromatography.

A series of α,β-unsaturated carboxylic ester substituted acetoacetates **130** were then subjected to the optimized reaction conditions (Table 6, entry 3). The reaction gave α-diazocarbonyl products as yellow oils in moderate yields (Table 7). Except for **130da** (Table 7, entry 6), no significant differences in yields were found with different carboxylic ester substituents. It should be noted that the products need to be stored in the freezer due to the poor stability of diazo compounds.

4.2 [2.2]Paracyclophane-based Ligands in Metal-Catalyzed N–H Insertion Reactions

In the past decade, ZHOU *et al.* revealed that the copper complexes bearing a chiral spiro bisoxazoline ligand show excellent performance in X–H (X = N, O, B) insertion reactions in both reactivity and enantioselectivity.[50,54,57,58,110] The crystal structures they obtained with various anions (PF_6^-, ClO_4^-, BAr_F^-) clarified that the two nitrogen atoms from each oxazoline coordinate with a copper(I) atom in a *trans* orientation and the phenyl groups on the oxazoline ligands form a perfect C_2-symmetric chiral pocket around the copper center to induce the chirality (Figure 15, **A**).

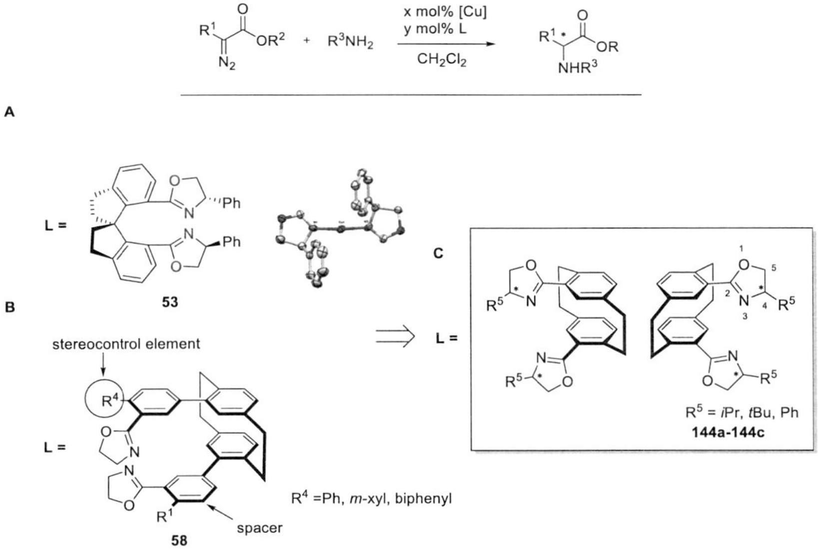

Figure 15. Design of novel pseudo-*ortho* [2.2]paracyclophane-based bisoxazoline ligands. **A**) Chiral spiro bisoxazoline ligand developed by ZHOU *et al.* and the corresponding chiral induction model. **B**) The pseudo-*ortho*-substituted aryl-[2.2]paracyclophane ligand scaffold developed by MUKAI. **C**) The designed pseudo-*ortho* [2.2]paracyclophane-based bisoxazoline ligands **144a–144c**.

In 2015, MUKAI *et al.* reported that the C_2-symmetric planar chiral [2.2]paracyclophane-based bisoxazoline ligands characterized by the inserted benzene as spacer. The spacer offered not only the conformational flexibility to make the distance between the two functional groups suitable for performing, but also a steric or electronic element based on the aryl group itself to interact with the substrates or reactants (Figure 15, **B**). In the copper-catalyzed intermolecular phenolic O–H insertion of α-diazoesters, the ligand achieved up to 80% ee.

Considering the high enantioselectivity both ligands achieved in metal-catalyzed O–H and N–H insertion reactions as well as the modifications of [2.2]paracyclophanes previously performed in our group, a series of pseudo-*ortho* [2.2]paracyclophane-based bisoxazoline ligands **144a–144c** (Figure 15, **C**) are designed in this work. The backbone would provide an inherent conformational rigidity,[62,150] and the direct connection of substituted chiral bisoxazoline on the backbone would offer a deep pocket to the induction of chirality. The ligands **144a–144c** can be used in copper-catalyzed N–H insertion reactions to generate the unsaturated α-amino carboxylic ester precursors **104**, which then can be transformed to the hexahydroindole motif **105** after a series of functionalization reactions.

4.2.1 Synthesis of Enantiopure pseudo-*ortho* [2.2]Paracyclophane-based Bisoxazoline Ligands

Chiral bisoxazoline ligands with high structural diversity have been introduced since 1989 and have received great attention in asymmetric catalysis.[103,151–156] The structures show particular stability in the presence of organolithium compounds, many reductive agents, weak acids, bases and GRIGNARD reagents. The conformational rigidity of the planar five-membered rings and the facile access to the chiral 4-position substituted derivatives have made them an important and versatile structural motif in enantioselective synthesis in organo- and metal-catalysis.[103,151,152] When the 4-substituted oxazoline ring coordinates with a metal *via* the nitrogen atom, the chiral information is brought very close to the metal center, making it particularly easy to transfer to the substrate, which results in high enantioselectivity.

The synthesis of the oxazoline ring can be achieved through condensation of the acid derivatives with the corresponding β-amino alcohol.[155,156] Therefore a condensation of β-amino alcohols **145a–145c** with the enantiopure pseudo-*ortho* [2.2]paracyclophane dicarboxylic acid **146** was chosen for the synthesis of enantiopure pseudo-*ortho* [2.2]paracyclophane-based bisoxazoline ligands. The enantiopure **146** can be obtained *via* lithiation-carboxylation[157] from an enantiopure pseudo-*ortho* dibromo[2.2]paracyclo-phane or *via* resolution from racemic dicarboxylic acid **146** by taking advantage of chiral auxiliary reagents.[158] Based on former experience, a chiral separation of (*rac*)-pseudo-*ortho* dibromo[2.2]paracyclophane **75** *via* semi-preparative HPLC (High Performance Liquid Chromatography) was used to access the enantiomer. Thus, the synthesis of (*rac*)-pseudo-*ortho* [2.2]paracyclophane **75** is needed (Scheme 27).

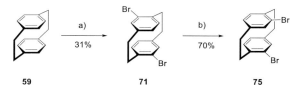

Scheme 27. Synthesis of pseudo-*ortho* dibromo[2.2]paracyclophane **75**. Reaction conditions: *a) Br₂ (2.20 equiv.), cat. Fe (0.05 equiv.), CCl₄, r.t., CH₂Cl₂, 3d, 31%; b) microwave, 180 °C, 17.2 bar, 300 W, 6 min, DMF, 4 times, 70%.*

Starting from the unsubstituted [2.2]paracyclophane **59**, iron-catalyzed dibromination[71,159] gave pseudo-*para* dibromo[2.2]paracyclophane isomer **71** as the major product, which can be isolated *via* filtration due to its bad solubility in dichloromethane and then recrystallized in hot toluene in 31% yield.[77,95,160,161] The microwave-assisted isomerization resulted in the smooth formation of pseudo-*ortho* dibromo[2.2]paracyclophane **75** in 70% yield after 4 cycles.[84] The ethylene bridge of **71** was cleaved by microwave irradiation and the resulting benzylic diradical underwent free rotation and recombined to generate a new ethylene bridge to form the new pseudo-*ortho* substitution.[84] CRAM *et al.* also showed that the pseudo-*para* dibromide **71** can be thermally isomerized into pseudo-*ortho* dibromide **75** by heating it to 200 °C in triglyme.[162] However, the microwave-assisted irradiation remarkably reduces the reaction time and dimethylforamide can be superheated until 180 °C to perform the reaction. The two enantiomers *(S*ₚ*)*-**75** and *(R*ₚ*)*-**75** can be separated *via* a semi-preparative CHIRALPAK® AZ-H column (Scheme 28). The two obtained fractions isolated at $t_R = 6.97$ min and $t_R = 8.00$ min show R_p and S_p configurations respectively.

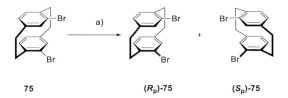

Scheme 28. Separation of racemic pseudo-*ortho* dibromide **75** *via* chiral semi-preparative HPLC. Conditions: *Semi-preparative CHIRALPAK® AZ-H column (20 × 250 nm, particle size 5 μm), 100% CH₃CN, 25 mL/min, 25 °C, 254 nm UV-detector, 100 mg racemate per run.*

Both obtained fractions were subjected to a lithiation-carboxylation protocol to generate the related [2.2]paracyclophane dicarboxylic acid (Scheme 29). (*R*ₚ)-pseudo-*ortho* [2.2]paracyclophane dicarboxylic acid (*R*ₚ)-**146** was obtained in 52% yield when *(R*ₚ*)*-**75** was treated with four equivalents of *tert*-butyllithium followed by subsequent carboxylation with pre-dried dry ice. However, a large amount of the *tert*-butyl keto substituted byproduct (S_p)-**147** was found when (S_p)-pseudo-*ortho* dibromo[2.2]para-cyclophane was treated with the same conditions. The formation of byproduct (S_p)-**147** can be accounted to the interaction between organolithium derivatives and lithium carboxylates present in the reaction mixture as well as their reactions with *tert*-butyllithium and carbon dioxide.[157,163]

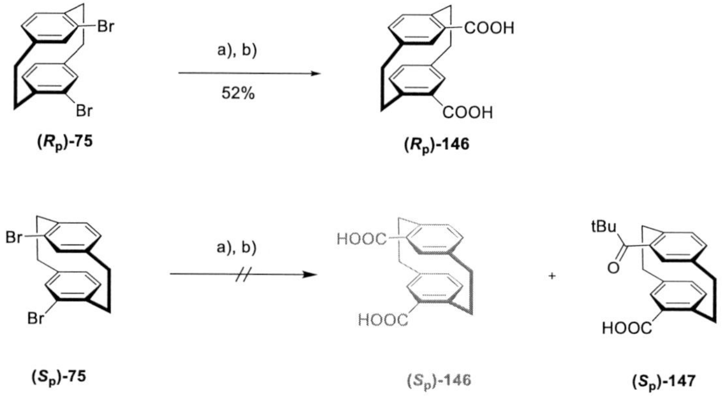

Scheme 29. Synthesis of (R_p)-4,12-dicarboxy[2.2]paracyclophane (R_p)-**146** and (S_p)-4,12-dicarboxy-[2.2]paracyclophane (S_p)-**146**. Reaction conditions: *a) tBuLi (4.00 equiv.), THF, −78 °C to r.t., 2 h; b) dry CO$_2$.*

The synthesis of (R_p)-pseudo-*ortho* [2.2]paracyclophane-based bisoxazoline derivatives can be achieved from the (R_p)-dicarboxylic acid (R_p)-**146**, which is firstly converted to the carboxylic acid chloride *in situ* and subsequently condensed with the (*S*)-valinol **145a**, (*S*)-*tert*-leucinol **145b** and (*S*)-phenylglycinol **145c**, respectively. The hydroxyl amides **148a**–**148c** are subsequently cyclized and dehydrated under APPEL conditions[164] affording the novel *iso*-propyl- (→**144a**), *tert*-butyl- (→**144b**) and phenyl-substituted (→**114c**) (R_p)-pseudo-*ortho* [2.2]paracyclophane-based bisoxazoline ligands (R_p)-**144a**–**144c** in 42%, 74% and 95% yields, respectively.[100,165]

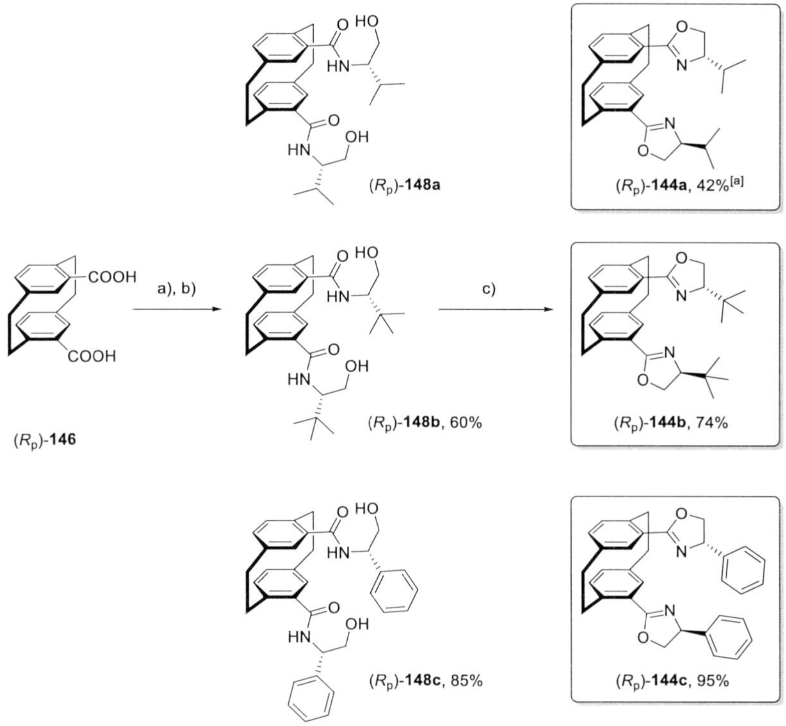

Scheme 30. Synthesis of pseudo-*ortho* [2.2]paracyclophane-based bisoxazoline ligands (*R*$_p$)-**144a–144c**. Reaction conditions: *a) SOCl$_2$, reflux, 3 h; b), β-amino alcohols* **145a–145c**, *Et$_3$N, CH$_2$Cl$_2$, r.t., overnight; c) Et$_3$N, PPh$_3$, CCl$_4$, CH$_3$CN, r.t., overnight.* [a] The reaction was carried out without the purification of the hydroxyl amide intermediate.

4.2.2 Synthesis of Ruthenium Complexes with [2.2]Paracyclophane Oxazoline Ligands

Ruthenium(II) complexes prefer monometallic structures and are generally stable under air and moisture with η^6-coordinated *p*-cymene as a substituent. The synthesis of cyclometalated ruthenium complexes of 2-phenyloxazoline has already been reported *via* an acetate-assisted C–H bond activation/deprotonation with the participation of [RuCl$_2$(*p*-cymene)]$_2$ precursor.[166–169] In 2017, BRÄSE *et al.* reported the cycloruthenation reaction of the [RuCl$_2$(*p*-cymene)]$_2$ precursor with [2.2]paracyclophane *N*-donor ligands, *i.e.* pyridyl, pyrimidyl and oxazolinyl (Scheme 31). By means of adding KPF$_6$ [170–172] the cationic complexes can be obtained in good yields.[173] Therefore, the 4-phenyloxazoline-substituted [2.2]paracyclophane **155** should be converted to the corresponding ruthenium complex under the same conditions.

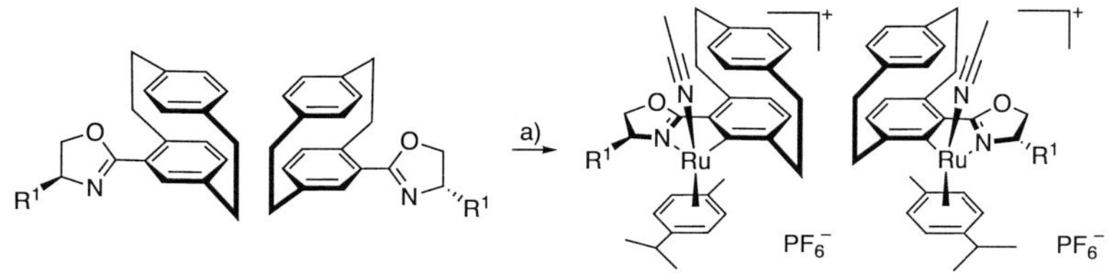

R^1 = *i*Pr, (R$_p$,S)/(S$_p$,S)-**149a**
R^1 = *t*Bu, (R$_p$,S)/(S$_p$,S)-**149b**

(R,S$_p$,S)-**150a**/(S,R$_p$,S)-**150a**, 52%
(R,S$_p$,S)-**150b**, 71%

Scheme 31. Cycloruthenation of [RuCl$_2$(*p*-cymene)]$_2$ with [2.2]paracyclophane oxazoline ligands. R^1 = *i*Pr, (R, S$_p$, S)-**150a**/(S, R$_p$, S)-**150b** were obtained as a mixture in a ratio of 1.18:1; R^1 = *t*Bu, (S, R$_p$, S)-**150b** is disfavored. Conditions: *a) [RuCl$_2$(p-cymene)]$_2$ (0.50 equiv.), KOAc (1.50 equiv.), KPF$_6$ (2.00 equiv.), CH$_3$CN, r.t., 3 d.*

The 4-phenyloxazoline-substituted [2.2]paracyclophane **155** can be obtained from the unsubstituted [2.2]paracyclophane **59** *via* a four-step synthesis (Scheme 32).

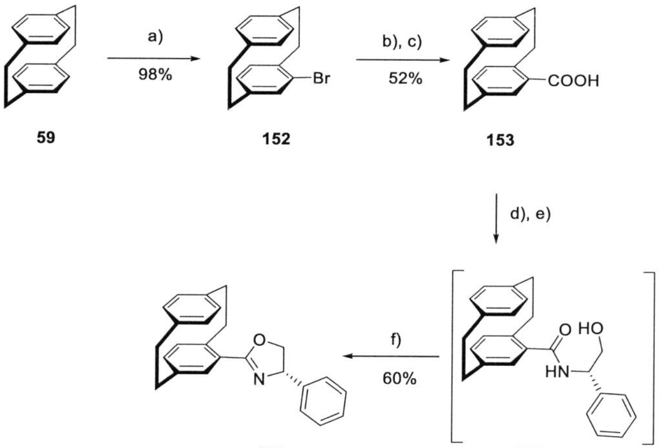

Scheme 32. Synthesis of 4-phenyloxazoline-substituted [2.2]paracyclophane ligand **155**. Reaction conditions: *a) Br$_2$ (1.00 equiv.), cat. Fe (0.05 equiv.), CCl$_4$, r.t., CH$_2$Cl$_2$, overnight; b) nBuLi (1.60 equiv.), THF, −78 °C to r.t.; c) CO$_2$; d) SOCl$_2$, 60 °C, 4h; e) (S)-(+)-phenylglylcinol (2.00 equiv.), Et$_3$N (2.00 equiv.), CH$_2$Cl$_2$; r.t., overnight; f) Et$_3$N (1.75 equiv.), PPh$_3$ (1.75 equiv.), CCl$_4$ (1.75 equiv.), CH$_3$CN, r.t., overnight.*

The mono-substituted bromide **152** was obtained *via* an iron-catalyzed bromination with one equivalent of bromide.[71–74,76,81,174,175] After a halogen-lithium exchange with *n*-butyllithium and carboxylation with carbon dioxide, the carboxylic acid **153** was generated as off-white solid,[176] which was then treated with thionyl chloride and subsequently condensed with (S)-(+)-phenylglycinol 145c to give the hydroxyl amide 154. Without further purification, the hydroxyl amide 154 was cyclized and dehydrated under APPEL conditions to afford 155 in 60% yield.[100,177] Since the [2.2]paracyclophane scaffold is racemic, the condensation product 155 also occurs as a mixture of diastereomers.

The obtained diastereomers (S_p,S)-155 and (R_p,S)-155 were then subjected to the cycloruthenation reaction as a mixture (Scheme 33). The fraction (R,S_p,S)-155 was considered to react analogous to the observed reactivity in a diastereoselective manner as (R_p,S)-149b (Scheme 31), so that only one of the diastereomeric products (R,S_p,S)-151 would be formed (Scheme 33). Indeed, the cycloruthenation product was obtained in a ratio of 1:1 without diastereoselectivity according to NMR signal integration. This can be explained by the appearance of a planar phenyl ring in comparison to the bulky *tert*-butyl group that is able to avoid steric hindrance by rotation. The obtained ruthenium complexes (R,S_p,S)-151 and (S,R_p,S)-151 showed sufficient stability for purification *via* column chromatography and the exact mass can still be found after being stored in air and moisture after a few weeks, although the color changed from yellow to green during this time.

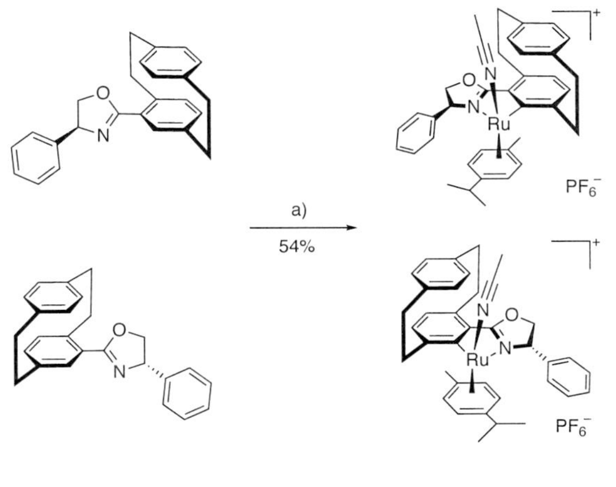

$(R_p,S)/(S_p,S)$-155 (R,S_p,S)-151/(S,R_p,S)-151

Scheme 33. Synthesis of the ruthenium complexes (R,S_p,S)-151/(S,R_p,S)-151 starting from 4-phenyloxazoline-substituted [2.2]paracyclophane $(S_p,S)/(R_p,S)$-155 as a diastereomer. Reaction condition: a) [RuCl₂(p-cymene)]₂ (0.50 equiv.), KOAc (1.50 equiv.), KPF₆ (2.00 equiv.), CH₃CN, r.t., 3 d.

Recently, IWASA *et al.* reported that the Ru(II)-phenyloxazoline complexes **156** [Ru(II)-Pheox] can be used as an effective catalyst in the enantioselective cyclopropanations,[178–184] Si–H insertion[185] as well as N–H insertion reactions.[186] These catalysts are easy to handle and stable in air and moisture.[179] Moreover, the Ru(II)-Pheox catalysts have a stereodirecting unit (blue) directly attached to the oxazoline ring and the electron density on the metal center can be controlled by featuring various substituents on the ligand backbone (red) (Figure 16).

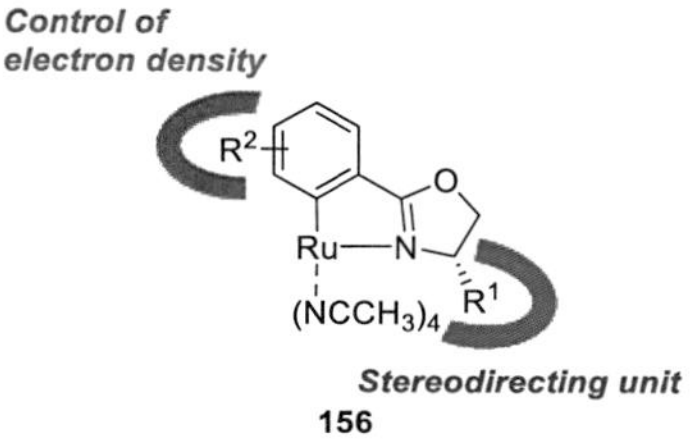

Figure 16. Modification of Ru(II)-Pheox **156**. R^1 = *i*Pr, *t*Bu, Ph, Bn. R^2 = H, MeO, Me₂N, Cl, CH₂OH.

By introduction of [2.2]paracyclophane into the backbone of the Ru(II)-Pheox complex, the electron density on the metal center can be changed, which can consequently influence the reaction activity. The inseparable diastereomeric mixture of (S_p,S)-**155** and (R_p,S)-**155** is subjected to the conditions reported by JUTAND *et al.* (Scheme 34).[187]

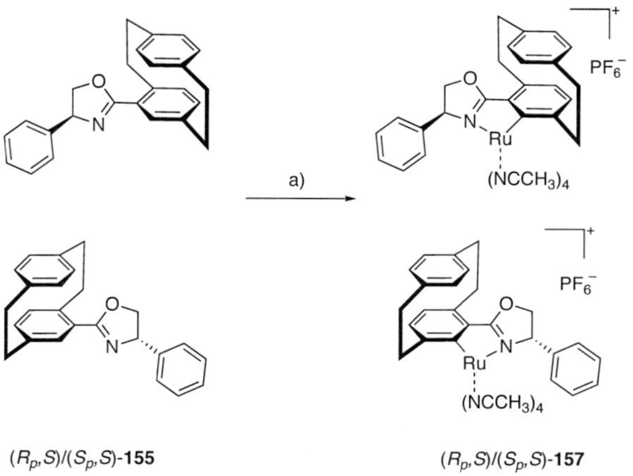

Scheme 34. Synthesis of ruthenium(II)-complexes **157** starting from the 4-phenyloxazoline-substituted [2.2]paracyclophane $(S_p,S)/(R_p,S)$-**155**. Reaction conditions: *a)* [RuCl₂Ph]₂ (0.50 equiv.), KOAc (1.50 equiv.), KPF₆ (2.00 equiv.), CH₃CN, 80 °C, 80 h.

After stirring the reaction mixture at 80 °C for 3 days, an aliquot from the mixture was taken out, diluted with deuterated acetonitrile and then characterized via ^{1}H NMR spectroscopy. The reaction showed not full conversion even after another 24 hours of stirring. Therefore, flash chromatography was applied to purify the mixture. The yellow liquid obtained after purification turned dark while concentrating under reduced pressure and decomposed during the ^{13}C NMR measurement. A technique to purify the product under argon atmosphere would be needed to obtain the Ru(II)-complex **157**. For the further study in catalyst screening, the (R_p,S)/(S_p,S)-**157** was generated in situ.

4.2.3 Synthesis of Amino-substituted [2.2]Paracyclophane Derivatives

Among the various disubstituted [2.2]paracyclophanes, P,N-bidentate ligands are often more stable than P,P-bidentate ligands and are amenable to the introduction of additional stereochemistry through the nitrogen moiety.

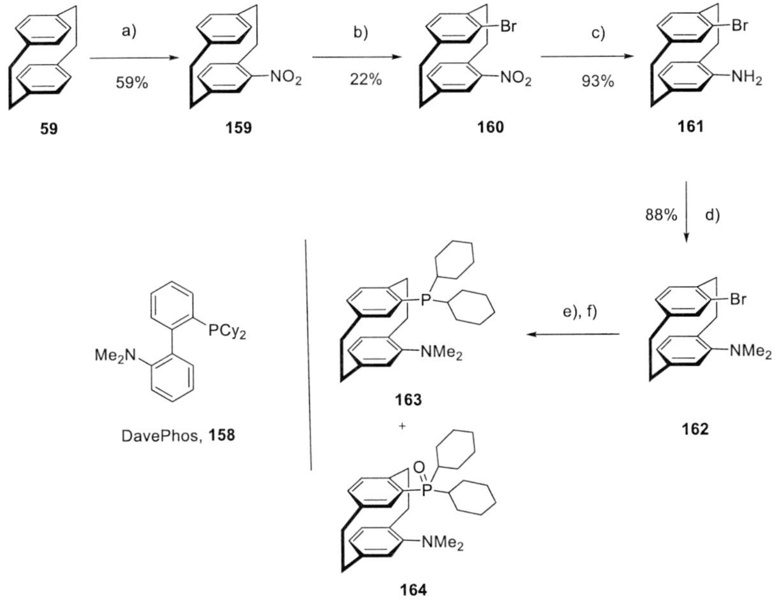

Scheme 35. Synthesis of DAVEPHOS-like P,N-bidentate [2.2]paracyclophane ligand. Reaction conditions: a) fuming HNO₃ (2.00 equiv.), glacial AcOH, 60 °C, 1 min; b) Br₂ (1.00 equiv.), cat. Fe (0.05 equiv.), CCl₄, r.t., CH₂Cl₂, overnight; c) Fe (12.0 equiv.), conc. HCl, EtOH/H₂O = 1:1 (v/v); d) MeI, K₂CO₃, DMF, 50 °C, overnight; e) nBuLi (1.10 equiv.);-78 °C, THF, 1 h; f) Cy₂PCl (1.40 equiv.), r.t., 6h.

Inspired from the widely spread application of DAVEPHOS **158** in the BUCHWALD-HARTWIG C–N cross-coupling,[1,188] α-arylation,[189] borylation[190,191] and other examples,[192,193] we have been interested in the introduction of *N,N*-dimethyl and dicyclohexylphosphine to the [2.2]paracyclophane backbone (Scheme 35).

The 4-nitro[2.2]paracyclophane **159** can be obtained *via* nitration by treating starting material unsubstituted [2.2]paracyclophane **59** in glacial acetic acid with fuming HNO_3 at a temperature not exceeding 60 °C.[77,194] Dinitro-substituted [2.2]paracyclophane was found when HNO_3 was added to the mixture at higher temperature. An iron-catalyzed bromination gave product **160** in 22% yield. After a Fe/HCl reduction and subsequent methylation, 13,13-dimethylamino-4-bromo[2.2]para-cyclophane **162** was obtained in 88% yield as a white solid. The crystal structures of **159**, **160** and **162** are shown in Figure 17. Afterwards, **162** is subjected to a halogen-lithium exchange with *n*-butyllithium followed by the addition of chlorodicyclohexylphosphine, which resulted in the smooth formation of a C–P bond. However, the product is not stable during flash chromatography as the oxidized product **164** was found from both NMR spectroscopy and mass spectrometry. A technique to purify the product under argon atmosphere or a protocol proceeds the full conversation of starting material is needed.

159   161   162

Figure 17. Crystal structures of 4-nitro[2.2]paracyclophane **159**, 13-amino-4-bromo[2.2]paracyclophane **160** and 13,13-dimethylamino-4-bromo[2.2]paracyclophane **162**, respectively.

As the bromination step in Scheme 35 gave product **160** in only 22% yield, another route aimed to improve the yield was carried out. Starting from **152**, the 4-amino[2,2]paracyclophane **161a** was obtained in 47% yield after lithiation, transformed to the corresponding azide with 4-acetamidobenzenesulfonyl azide (*p*-ABSA) and reduction with sodium borohydride by adding absolute methanol dropwise. Following the procedure shown in Scheme 36, the product 4,4-dimethylamino[2.2]paracyclophane **162a** was obtained in 95% yield and then subjected to the iron-catalyzed bromination. However, the reaction showed no transformation from the starting material after stirring overnight.

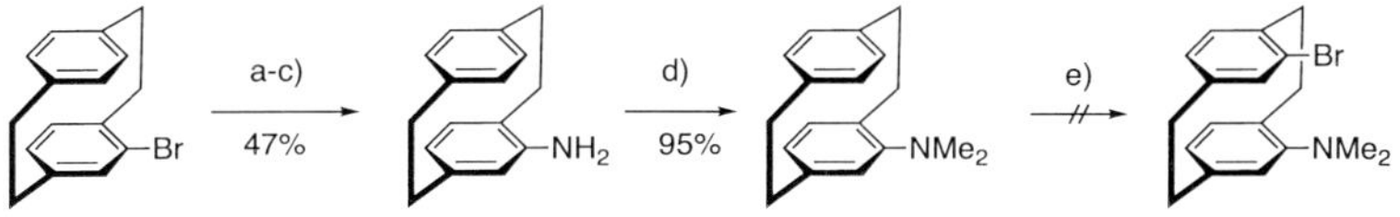

152 161a 162a 162

Scheme 36. Synthesis of (*rac*)-13,13-dimethylamino-4-bromo[2.2]paracyclophane **162**. Reaction conditions: *a) nBuLi (1.15 equiv.), THF,–78°C to r.t., 1 h; b) p-ABSA (1.15 equiv.), THF; c) $NaBH_4$ (10.0 equiv.), abs. MeOH (dropwise), reflux, 5 h; c) MeI, K_2CO_3, DMF, 50 °C, overnight; e) Br_2 (1.00 equiv.), cat. Fe (0.05 equiv.), CCl_4, r.t., CH_2Cl_2, overnight.*

This can be actually accounted for the *N,N*-dimethyl substituent. Pseudo-*gem* substitution is sterically unfavored. The first introduced monosubstituent must be a LEWIS basic group, which could capture the pseudo-*gem* proton *via* an intramolecular deprotonation process to make the next substitution on pseudo-*gem* position kinetically favorable.

4.3 Metal-catalyzed N–H Insertion with [2.2]Paracyclophane Ligands

4.3.1 Copper-Catalyzed N–H Insertion of Saturated α-Diazocarbonyl Compounds with [2.2]Paracyclophane Ligands

According to the results published by ZHOU *et al.*, dichloromethane is proven to be the best solvent for metal-catalyzed N–H insertion reactions[50,54,57] and Cu(I) is the catalytically active oxidation state in copper carbenoid chemistry.[31] Therefore, this study focuses on searching for the best copper(I) catalyst precursor under the reported optimized conditions.

Table 8. Copper-catalyzed insertion of benzyl-2-diazopropionate **113d** into the N–H bond of aniline **165a** in the presence of [2.2]paracyclophane-based bisoxazoline ligands **144a**. [a]

Entry	[Cu]	T [°C]	Yield [%][b]	
			166	167
1	Cu(MeCN)$_4$PF$_6$	40	77	18
2	CuCl	40	-	88
3	[CuOTf]$_2$·Tol	40	58	38
4	Cu(MeCN)$_4$PF$_6$	r.t.	93	5
5	[CuOTf]$_2$·Tol	r.t.	61	22
6[c]	Cu(MeCN)$_4$PF$_6$	r.t.	13	84

[a] Reaction conditions: [Cu] (5.00 μmol), **144a** (6.00 μmol) and NaBARF (6.00 μmol) in 1 mL of CH$_2$Cl$_2$ was stirred at 40 °C for 2 h, then aniline (**165a**, 0.10 mmol) and benzyl-2-diazopropionate (**113d**, 0.12 mmol) were added subsequently and the mixture was stirred at r.t. for 30 min. [b] Yields were determined by GC-MS. [c] The reactions were carried out without the participation of ligand. - = no product.

Initially, the insertion of benzyl 2-diazopropionate **113d** into N–H bond of aniline **165a** was performed in dichloromethane at 40 °C with the catalyst generated *in situ* from 5 mol% of Cu(MeCN)$_4$PF$_6$ and 6 mol% of *(S_p,S)/(R_p,S)*-**144a**. The insertion product, benzyl α-phenyl-aminopropionate **166**, was obtained in 77% yield accomplished with 18% of β-elimination product (methyl cinnamate, **167**) (Table 8, entry 1). Various copper catalyst precursors, including CuOTf and CuCl were tested in the reaction (Table 8, entry 2–3). The nature of the counterions of the copper influenced the reactivity of the catalysts. The smaller and stronger coordinating OTf$^-$ ion is inferior to the PF$_6^-$ in the reactivity. The use of the neutral copper catalyst precursor CuCl gave no product. Decreasing the reaction temperature can slightly improve the yields of product (Table 8, Entry 4–5). Furthermore, we tested the reaction without the participation of ligand (Table 8, entry 6), the yield showed sharply diminish. The large difference in yields clearly clarified the importance of the ligand as well as the catalytic ability of [2.2]paracyclophane-based bisoxazoline ligand in the copper-catalyzed N–H insertion reaction. Since the ligand we used in the reaction is a mixture of S_p and R_p configuration, the enantioselectivity is not accessible.

The competitive β-elimination from the copper carbene intermediate has a large influence on the yield of the N–H insertion product. Although this can be suppressed by changing the counterions (93%, Table 8, entry 4), another study without the competition of β-elimination was performed with phenyl-2-diazopropionate **118** as starting material. Although the dimerization product, dimethyl 2,3-diphenylmaleate **169** was found after the reaction, this can be actually avoided by adding the starting material **118** slowly into the *in situ* generated catalyst at room temperature.

As we expected, without the competition of the β-elimination, the reaction showed better yields in generating phenyl α-phenyl-aminopropionate **168**. Similarly, PF$_6^-$ proved to be the best counterion and gave the product in 98% yield (Table 9, entry 2). Besides Cu(MeCN)$_4$PF$_6$, CuCl (Table 9, entry 1) and CuOTf (Table 9, entry 3) showed 44% and 64% yields, respectively. Without the participation of ligand, the product was obtained in 40% yield using Cu(MeCN)$_4$PF$_6$ as copper source.

Table 9. Copper-catalyzed insertion of phenyl-2-diazopropionate **118** into the N–H bond of aniline **165a** in the presence of [2.2]paracyclophane-based bisoxazoline ligand **144a**.[a]

		118	165a	166	$(S_p, S)/(R_p, S)$-**144a**

167

Entry	[Cu]	Yield [%][b]	
		168	169
1	CuCl	44	<2
2	Cu(MeCN)$_4$PF$_6$	98	2
4	[CuOTf]$_2$·Tol	64	28
5[c]	Cu(MeCN)$_4$PF$_6$	40	-

[a] Reaction conditions: [Cu] (5.00 μmol), **144a** (6.00 μmol) and NaBARF (6.00 μmol) in 1 mL of CH$_2$Cl$_2$ was stirred at 40 °C for 2 h, then aniline (**165a**, 0.10 mmol) and phenyl-2-diazopropionate (**118**, 0.12 mmol) were added subsequently and stirred at r.t. [b] Yields were determined by GC-MS. [c] The reaction was carried out without the participation of ligand.

Under the optimized condition, a variety of substituted amines were examined in the copper-[2.2]paracyclophane-catalyzed N–H insertion from saturated α-diazoesters (Table 10). All substituted anilines underwent the insertion reaction with high reactivity and complete conversions were achieved within 10 minutes. The corresponding insertion products were obtained in good to excellent yields regardless of the nature and the position of the substituents of the aniline derivatives (entry 1–10). Introduction of an electron-donating group on the *meta* position (entry 5) slightly diminished the yield. The aliphatic amine, cyclohexylamine, showed no reactivity under the identical reaction conditions (entry 11–12). Among all the obtained products, the crystal structure of phenyl α-phenyl-aminopropionate (**170b**, entry 2) was measured and shown in Figure 18.

Table 10. Copper-catalyzed insertion of saturated α-diazocarbonyl compounds into the N–H bond of anilines in the presence of [2.2]paracyclophane-based bisoxazoline ligands **144a**. [a]

$(S_p,S)(R_p,S)$-**144a**

Entry[a]	R^1	R^2	R^3	Product	Yield [%][b]
1	Bn	Me	Ph	**170a**	77
2	Ph	Me	Ph	**170b**	98
3	Me	Bn	Ph	**170c**	94
4	Me	Bn	o-MeOPh	**170d**	82
5	Me	Bn	m-MeOPh	**170e**	53
6	Me	Bn	p-MeOPh	**170f**	70
7	Me	Bn	o-MePh	**170g**	68
8	Me	Bn	p-MePh	**170h**	70
9	Me	Ph	Ph	**170i**	68
10	Me	tBu	Ph	**170j**	74
11	Me	Bn	c-C$_6$H$_{11}$	**170k**	-
12	Me	Ph	c-C$_6$H$_{11}$	**170l**	-

[a] Reaction conditions: Cu(MeCN)$_4$PF$_6$ (5.00 μmol), $(S_p,S)/(R_p,S)$-**144a** (6.00 μmol) and NaBARF (6.00 μmol) in 1 mL of CH$_2$Cl$_2$ was stirred at 40 °C for 2 h, then aniline (0.10 mmol) and α-diazocarbonyls (0.12 mmol) were added subsequently and stirred at r.t. [b] isolated yields. - = no product.

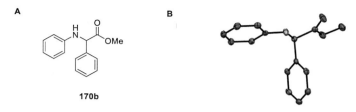

Figure 18. A) Methyl 2-phenyl-2-(phenylamino)acetate **170b**. **B)** Crystal structure of **170b**.

4.3.2   Copper-Catalyzed   N–H   Insertion   of   Unsaturated   α-Diazocarbonyl Compounds with [2.2]Paracyclophane Ligands

With the former experience in hand and the proof that copper catalysts with [2.2]paracyclophane ligands work well with saturated α-diazocarbonyl compounds, we started the research on the enantioselective N–H insertion reaction from unsaturated α-diazocarbonyl compounds. The aim was to generate the δ-amino-α,β-unsaturated carboxylic esters **104**, which can then be transformed to the hexahydroindole motif **105** *via* a series of functionalization reactions (Chapter 4.1.3).

We started our investigations with the comparison of two different Cu(I)-complexes to be sure about the reactivity and enantioselectivity of (S_p,S)-4,12-bis-(4'-isopropyloxazolin-2'-yl) [2.2]paracyclophane (S_p,S)-**144a**[III] in copper-catalyzed asymmetric N–H insertions. Aniline **165a** and 1-Ethyl 6-methyl (*E*)-5-diazohex-2-enedioate **103b** were chosen as a model system for the screening (Table 11). Only the β-elimination byproduct was found without the participation of ligand (Table 11, entry 1). To our delight, the insertion product **171** was obtained in 21% yield when the *in situ* generated catalyst from 5 mol% of Cu(MeCN)$_4$PF$_6$ and 6 mol% of ligand **144a** was part of the reaction (Table 11, entry 2). As a comparison, the bisoxazoline ligand (S,S)-Ph-Pybox **172** yielded 25% of N–H insertion product (Table 11, entry 3). The yield was improved to 38% when twice amount of the copper-[2.2]paracyclophane-based bisoxazoline catalyst was added to the reaction (Table 11, entry 4).

[III] The ligand was synthesized by Dr. Carolin Braun.

Table 11. Initial screening of ligands in the copper-catalyzed insertion of **103b** into N–H bond of aniline **165a**. [a]

	(S, S)-**172**	(R_p, S)-**144a**	

Entry	x [mol%] Ligand	y [mol%] Cu[Me(CN)₄]PF₆	Yield [mol%][c]
1[b]	-	6	-
2	5 (144a)	6	22
3	5 (172)	6	25
4	10 (144a)	12	38

[a] Reaction conditions: Cu(MeCN)₄PF₆ (5.00 μmol), ligand (6.00 μmol) and NaBARF (6.00 μmol) in 1 mL of CH₂Cl₂ was stirred at 40 °C for 2 h, then aniline (**165a**, 0.10 mmol) and **103b** (0.12 mmol) were added subsequently and stirred at r.t. [b] Reaction was carried without the participation of ligand. [c] Isolated yields. - = no product.

The low yields on the one hand have proved that the copper-[2.2]paracyclophane-based bisoxazoline catalyst can in general work in N–H insertions with unsaturated α-diazocarbonyl compounds. On the other hand, it encouraged us to modify the ligand to find a more efficient catalyst for this reaction. Therefore, three novel (R_p,S)-pseudo-*ortho* [2.2]paracyclophane-based bisoxazoline ligands (R_p,S)-**144a**–**144c** were designed and synthesized *via* dibromination, microwave-assisted isomerization, semi-preparative HPLC-assisted separation, lithiation-carboxylation and condensation (Chapter 4.2.1). Moreover, two ruthenium complexes with oxazoline-substituted [2.2]paracyclophane ligand (Chapter 4.2.2) were also used in the further catalyst screening.

The effects of different catalysts on the yields and ee values in the insertion of unsaturated α-diazocarbonyl compound **103e** into aniline **165a** are summarized in Table 12. To our delight, the yields have been improved by using the *in situ* generated catalyst with the designed new ligands (Table 12, entry 3–5). Ruthenium(II)-Pheox **156** showed good reactivity and the product **176d** was obtained in 74% yield (Table 12, entry 5). $(S_p,S)/(R_p,S)$-**157**, designed as a modification of ruthenium(II)-Pheox **156**, yielded 55% product (Table 12, entry 8). The catalyst $(S_p,S)/(R_p,S)$-**157** was generated *in situ* in absolute acetonitrile. Considering the negative effect of polar solvents on the reaction, $(S_p,S)/(R_p,S)$-**157** was pre-dried under vacuum and re-dissolved in degassed dichloromethane before aniline **165a** and the **103e** were added. The other ruthenium-complex $(S_p,S)/(R_p,S)$-**151** showed no activity under the identical conditions (Table 12, entry 6), but 22% yield of the product were obtained after the reaction was stirred at 40 °C overnight.

Table 12. Catalyst screening for the metal-catalyzed insertion of **103e** into N–H bond of aniline **165a**.

Entry	Catalyst	Yield [%][e]	ee [%][f]
1[a]	(S,S)-**172** + [Cu]	50	*rac*
2[a]	(R$_p$,S)-**144a** + [Cu]	38	*rac*
3[a]	(R$_p$,S)-**144b** + [Cu]	80	*rac*
4[a]	(R$_p$,S)-**144c** + [Cu]	88	*rac*
5[b]	**156**	74	*rac*
6[b]	(S$_p$,S)/(R$_p$,S)-**151**	-	-[g]
7[b][c]	(S$_p$,S)/(R$_p$,S)-**151**	22	-
8[d]	(S$_p$,S)/(R$_p$,S)-**157**	55	-

[a] Reaction conditions: Cu(MeCN)$_4$PF$_6$ (10.0 µmol), ligand (12.0 µmol) and NaBARF (12.0 µmol) were stirred in 1 mL of CH$_2$Cl$_2$ at 40 °C for 2 h, then aniline (**165a**, 0.10 mmol) and **103e** (0.12 mmol) were added subsequently and stirred at r.t. [b] Ruthenium-complex (5 mol%) was dissolved in 1 mL of CH$_2$Cl$_2$, aniline (**165a**, 0.10 mmol) and **103e** (0.12 mmol) were added subsequently and stirred at r.t for 2 h. [c] Reaction temperature: 40 °C. [d] Metal complex was generated *in situ*. [e] Isolated yields. [f] Determined by chiral HPLC on a CHIRALPAK AS column. [g] The used ruthenium complexes are S$_p$/R$_p$ mixtures, no ee values are accessible.

In terms of reactivity, the pseudo-*ortho* [2.2]paracyclophane-based bisoxazoline copper complexes are superior to the mono-substituted bisoxazoline ruthenium complexes as well as the Ph-Pybox (S,S')-**172** copper complex and the Ru(II)-Pheox **156**. In order to get deeper understanding of the enantioselectivity of the different catalysts, the ee values were determined by chiral HPLC on a CHIRALPAK AS column [IV]. However, all the N–H insertion products obtained from chiral catalysts (S,S)-**172** and (R$_p$,S)-**144a**–**144c** showed no enantiomeric excess.

[IV] HPLC condition: CHIRALPAK AS column, n-hexane/2-propanol = 85:15, flow rate = 1.0 mL/min, 15 °C, λ = 254 nm, t_{R1} = 18.79, t_{R2} = 28.04.

For the sake of understanding the catalyst structure, we carried out the reaction of (S_p,S)-**144a** with Cu(MeCN)$_4$PF$_6$ in absolute dichloromethane to analyze the *in situ* generated catalyst. After stirring under 40 °C for 2 h, an aliquot from the reaction mixture was taken out, dried under vacuum, diluted into 500 μL of deuterated chloroform and characterized by ^{1}H NMR spectroscopy. The result revealed the presence of a monomeric coordination in the solution according to the NMR shift on $H^{5'}$ and $H^{4'}$ (Figure 19). Matrix-assisted laser desorption ionization-time of flight mass spectrometry (MALDI-ToF-MS) also confirmed the monomeric species in the solution (m/z = 494, {Cu[(S_p,S)-144a]}$^+$).[V]

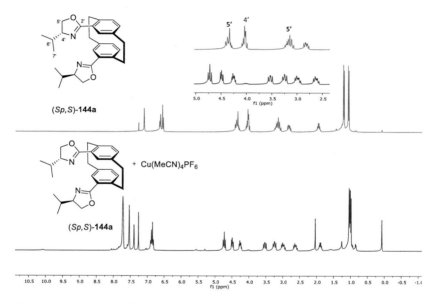

Figure 19. Comparison of the ^{1}H NMR spectra of the ligand **144a** and the catalyst generated *in situ* from **144a** and Cu(MeCN)$_4$PF$_6$ (300 MHz, CDCl$_3$).

Meanwhile, we attempted to understand the catalyst structure of Cu-(S_p,S)-**144a** by means of X-ray diffraction analysis. Through slow evaporation of dichloromethane, a white crystal of "*in situ* generated Cu-(S_p,S)-**144a** complex" was obtained. The crystallographic data showed only the characterization of the ligand (S_p,S)-**144a**, instead of the expected Cu-(S_p,S)-**144a** complex (Figure 20). However, it provided us an overview of the spatial conformation of the novel pseudo-*ortho* [2.2]paracyclophane-based bisoxazoline ligand (S_p,S)-**144a** and the distance between the two nitrogen atoms, which can be used as a comparison to other ligands, especially the spiro bisoxazoline ligands, to study the effect

---

[V] MS (MALDI-TOF, Matrix: CHA, +0.1% TFA): m/z = 431 [(S_p,S)-**144a**+H]$^+$, 494 {Cu[(S_p,S)-**144**]}$^+$.

of ligands on the results of asymmetric insertion of α-diazocarbonyl compounds into N–H bonds of aniline derivaties.

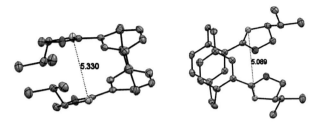

Figure 20. X-ray diffraction structures of ligand (S_p,S)-**144a**.

The spiro bisoxazoline ligands developed by ZHOU *et al.* provided excellent reactivity and enantioselectivity in N–H insertion reactions.[4,50,54,58] Their catalyst crystal structure studies revealed that Cu(I)-spiro bisoxazoline complex (Figure 21, **B**) has a binuclear structure, in which each of the two copper(I) atoms is coordinated by two nitrogen atoms from the two spiro-bisoxazoline ligands in a *trans* orientation.[58] And the phenyl groups on the bisoxazoline form a perfect C_2-symmetirc chiral pocket around the copper center. The minimized 3D structures of the spiro bisoxazoline ligand **53** presents a surprising picture. In the minimal energy conformer (Figure 21, **A**) the nitrogen atoms prefer to be more than 6 Å apart, which is precluding the bidentate chelation. But in the rotamer that would be required for copper chelation, there is only 3.0 Å between the two nitrogen atoms (Figure 21, **C**).[39]

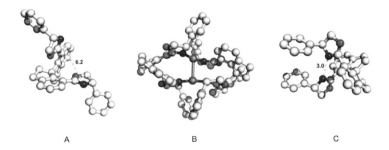

Figure 21. A) Lowest energy conformer of the spiro bisoxazoline (MM2 minimization). **B)** Crystal structure of the bimetallic complex determined by ZHOU *et al.*[58] **C)** Oxazole rotamer that would be required for a bidentate chelation with copper. Copyright© 2013 Royal Society of Chemistry.

Obviously, to meet the required conformation in which each copper atom coordinates with the two nitrogen atoms from the different ligands in a *trans* orientation, the conformation of the spiro bisoxazoline ligand **53** has been adjusted from the lowest energy conformer to a distance around 3.0 Å between two nitrogen atoms. It is possible that the copper-catalyzed N–H insertion reactions in the presence of Cu(I)-(S_p,S)-**144** showed no enantioselectivity due to the inherent conformational rigidity of the ligand backbone. The energy under room temperature is not enough for [2.2]paracyclophane to change its conformation to meet the required cooper chelation.

Furthermore the crystallographic data (Figure 20) shows that the distance between the two nitrogen atoms from the [2.2]paracyclophane-based bisoxazoline ligand (S_p,S)-**144a** is 5.330 Å and 5.089 Å, which means that the copper carbene (Figure 22) cannot be fixed in the 'pocket' generated by the ligand. Thus, the lone pair electrons from the nitrogen atom on the aniline could attack the carbene without selectivity to generate the ylide **174**.

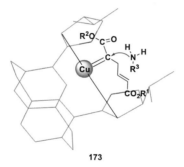

173

R^1 = alkyl, Bn, R^2 = alkyl, Bn.

Figure 22. Proposed model for the generation of the copper-associated ylide by the attack of amine N atom on the [2.2]paracyclophane-Cu carbene.

The generally accepted N–H insertion reactions proceed *via* a stepwise mechanism (Figure 23).[40,127,195,196] The copper-associated ylide **174** is formed by the attack of the lone pair electrons of the aniline N atom on the electron-deficient copper carbene. The ylide **174** may undergo one of the possible processes to generate the N–H insertion product. One process involves simultaneous proton transfer and catalyst dissociation (step c), the other involves dissociation of the copper catalyst to form a free ylide **175** (step d). Since the retention of the configuration of the free ylide **175** is difficult, the enantioselectivity cannot be assured in the proton transfer step (step d). Considering the conformation of (S_p, S)-**144** as well as the flexible N–Cu bond, it is reasonable to assume that the Cu(I)-(S_p,S)-**144** catalyzed N–H insertion proceeds *via* the pathway a → b → d → e. Moreover, the double bond of the unsaturated α-diazocarbonyl compound **103** provides an uncertain factor for the retention of the configuration of the free ylide **175**.

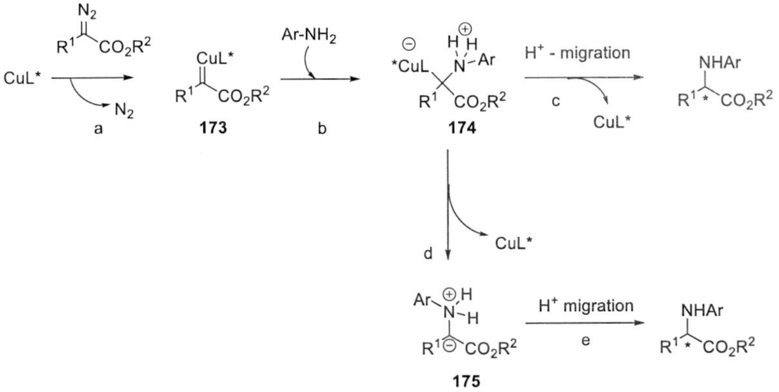

Figure 23. Proposed mechanism for copper-catalyzed insertion of saturated α-diazocarbonyl compounds into the N–H Bond of aniline, R^1 = alkyl, Bn, R^2 = alkyl, Bn.

Although the Cu-(S_p,S)-**144c** catalyst yielded 88% of product **176d** in the insertion of **103e** into the N–H bond of aniline **165a** (Table 12, entry 4), the complexity of the synthesis of pseudo-*ortho* [2.2]paracyclophane-based bisoxazoline ligands and the cost made us choose another ligand. Therefore, the commercially available (S,S)-Ph-Pybox **172** were used to generate unsaturated α-amino carboxylic esters **104** and to study the influence of substrates in the catalytic system, which may build a solid foundation for further research.

The influence of the α-diazoester side chain (R^1) and the unsaturated ester side chain (R^2) of the diazo compounds on the reactivity of the N–H insertion have been investigated. Five diazo compounds with different substituents (R^1, R^2) were chosen. The α-diazo-*tert*-butyl ester (Table 13, entry 5) yielded 75% product and the corresponding crystal structure is proven and shown in Figure 24. When the *tert*-butyl substitution was changed from R^1 to R^2 on the unsaturated ester side chain, the α-diazo-methyl ester gave the corresponding product in 39% yield (Table 13, entry 2). The decrease of yield can be accounted to the steric hinderance of the α-diazoester side chain (R^1), on which the sterically hindered *tert*-butyl substitution could better avoid the β-elimination of the *in situ* generated copper-carbene to improve the yield of the N–H insertion product. The other diazo compounds provided corresponding products in 42–60% yield (Table 13, entries 1, 3, 4).

Table 13. Cu-catalyzed insertion of unsaturated α-diazocarbonyl compounds **103** into N–H bond of aniline **165a** in the presence of (S,S)-Ph-Pybox **172**: the influence of α-diazocarbonyl compounds on yields.[a]

Entry	R^1	R^2	Yield [%][b]
1	Me	Et	49 (**176b**)
2	Me	*t*Bu	39 (**176c**)
3	Me	Bn	60 (**176d**)
4	Bn	Me	42 (**176e**)
5	*t*Bu	Me	75 (**176f**)

[a]Reaction conditions: Cu(MeCN)$_4$PF$_6$ (5.00 μmol), (S,S)-**172** (6.00 μmol) and NaBARF (6.00 μmol) in 1 mL of CH$_2$Cl$_2$ were stirred at 40 °C for 2 h, then aniline (**165a**, 0.1 mmol) and α-diazocarbonyls **103** (0.12 mmol) were added subsequently and stirred at r.t for 2 h. [b] Isolated yields.

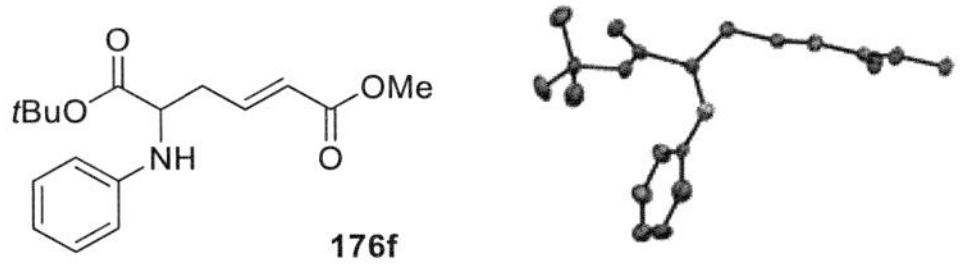

Figure 24. A) Molecular structure of 176f. B) Crystal structure of 176f.

After the initial screening with aniline 165a, we tried to apply the N–H insertion to different substituted anilines using Cu-(S,S)-172 as a catalyst. First, we used dimethyl (E)-5-diazohex-2-enedioate 103a as substrate and 4-phenyl- (165b), 4-methoxy- (165c) and 2-methoxyaniline (165d) as reagents, which gave the related products in 46%, 61% and 42% yields, respectively (Table 14, entries 1–3). The same anilines gave the corresponding products in slightly lower yields when the diazo substrate was changed to 6-benzyl-1-methyl-(E)-5-diazohex-2-enedioate 103f (Table 14, entries 4–6). The electronic properties of the anilines were tested by varying the substituents. Compared to 2-chloroaniline 165e, 2-iodoaniline 165f yielded just traces of product (Table 14, entries 7–8), 3,5-ditrifluroro-aniline 165g gave product in 40% yield (Table 14, entry 9). No product was found when 2,4-dichloro-6-iodo-aniline 165h was used in the reaction (Table 14, entry 10). The 3,5-dimethoxylaniline 165i sharply diminished the yield to 17% compared to 2-methoxyl-aniline 165d (Table 14, entry 11). The results demonstrated that the electronic property has a large influence on the reactivity: strong electron withdrawing or donating groups on the aniline lead to a decrease in yields. The 4-((trimethylsilyl)ethynyl)-aniline 165j gave the corresponding product in 70% yield (Table 14, entry 12), but as the reaction was catalyzed by Ru-Pheox 156, the results are not comparable. When we kept the electronic features of aniline but increased the steric bulk by using 3,5-di-tert-butylaniline 165k as a substrate (Table 14, entries 13, 15), the yield was slightly decreased.

Table 14. Cu-catalyzed insertion of unsaturated α-diazocarbonyl compounds 103 into N–H bonds of substituted anilines 165 in the presence of (S,S)-Ph-Pybox 172: the influence of aniline on yields.[a]

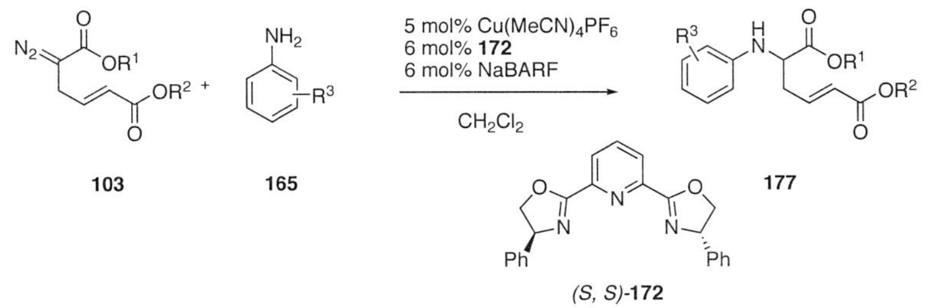

Entry[a]	R¹	R²	R³	Yield[b]
1	Me	Me	4-Ph (**165b**)	46 (**177ab**)
2	Me	Me	4-OMe (**165c**)	61 (**177ac**)
3	Me	Me	2-OMe (**165d**)	42 (**177ad**)
4	Bn	Me	4-OMe (**165c**)	43 (**177fc**)
5	Bn	Me	2-OMe (**165d**)	36 (**177fd**)
6	Bn	Me	4-Ph (**165b**)	36 (**177fb**)
7	Bn	Me	2-Cl (**165e**)	54 (**177fe**)
8	Bn	Me	2-I (**165f**)	Trace (**177ff**)
9	Bn	Me	3,5-di-CF₃ (**165g**)	40 (**177fg**)
10	Bn	Me	2,4-dichloro-6-iod (**165h**)	-[d] (**177fh**)
11	Bn	Me	3,5-di-OMe (**165i**)	17 (**177fi**)
12	Bn	Me	4-TMS (**165j**)	70[c] (**177fj**)
13	Bn	Me	3,5-di-*t*Bu (**165k**)	34 (**177fk**)
14	Me	Bn	4-OMe (**165c**)	44 (**177ec**)
15	Me	Bn	3,5-di-*t*Bu (**165k**)	41 (**177ek**)
16	Me	Et	4-Me (**165l**)	45 (**177bl**)

[a]Reaction conditions: Cu(MeCN)₄PF₆ (5.00 μmol), (*S,S*)-**172** (6.00 μmol) and NaBARF (6.00 μmol) in 1 mL of CH₂Cl₂ were stirred at 40 °C for 2 h, then α-diazocarbonyl compounds **103** (0.12 mmol) and substituted aniline **165** (0.10 mmol) were added subsequently and stirred at r.t. for 2 h. [b] Isolated yields. [c] The reaction was carried out by using Rh(II)-Pheox **156** as catalyst.

Besides aniline and the substituted anilines, the amino-substituted [2.2]paracyclophane derivatives **161** and **161a** can also react with the α-diazocarbonyl compounds to afford the insertion products. Although the yields are much lower (Scheme 37), it provided an alternative method to modify the amino-substituted [2.2]paracyclophane derivatives. The pyridine-2-amine and its analogues were completely inert under the identical reaction conditions.

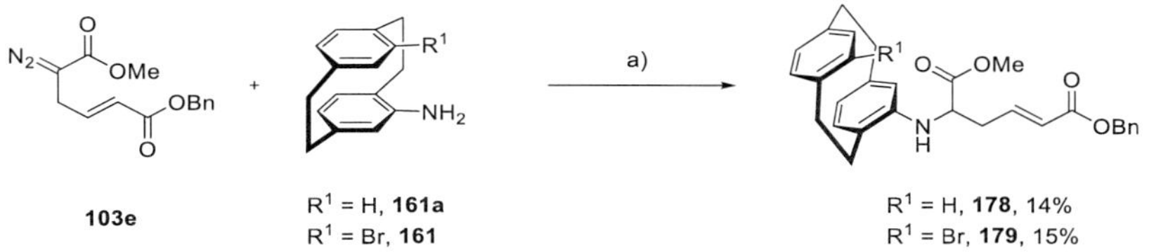

Scheme 37 Cu-(*S,S*)-**172**-catalyzed insertion of **103e** into N–H bond of amino-substituted [2.2]paracyclophane derivatives. Reaction condition: a) *Cu(MeCN)₄PF₆ (10.0 µmol), (S,S)-**172** (12.0 µmol) and NaBARF (12.0 µmol) in 1 mL of CH₂Cl₂ were stirred at 40 °C for 2 h, then **103e** (0.12 mmol) and amino-substituted [2.2]paracyclophane derivatives (0.10 mmol) were added subsequently and stirred at r.t. for 2 h.* [a]Isolated yields.

4.3.3 Application of the Copper-catalyzed N–H insertion in the Synthesis of Hexahydroindole Motif

With the experience of the N–H insertion reactions in hand, a route (Scheme 38) towards the synthesis of the hexahydroindole motif **105** of Rostratin B-D **106–108** was developed.

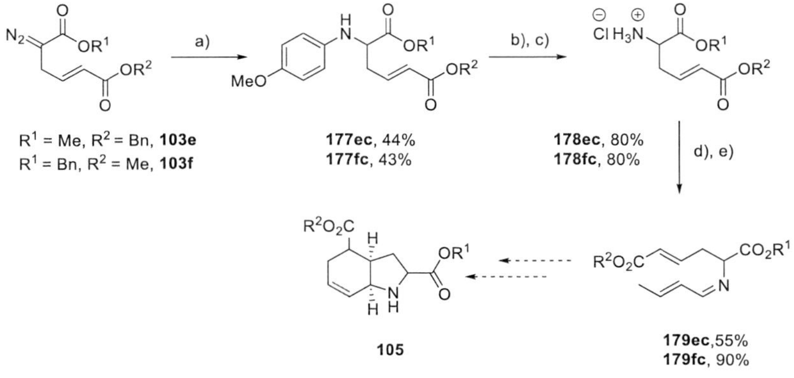

Scheme 38. Synthesis of the hexahydroindole motif **105** *via* a copper-catalyzed N–H insertion reaction. Reaction conditions: *a) p-anisidine (**165c**, 1.20 equiv.), CH₂Cl₂, r.t., 2 h, catalyst was generated in situ from 5 mol% of Cu(MeCN)₄PF₆, 6 mol% of (S,S)-**172** and 6 mol% of NaBARF in CH₂Cl₂; b) H₅IO₆, 1 M H₂SO₄, CH₃CN/H₂O = 1:1; c) HCl (1 M in EtOAc); d) Et₃N, CH₂Cl₂, r.t., 2 h; e) crotonaldehyde (1.10 equiv.), 4Å MS, Et₂O; r.t., overnight.*

The unsaturated α-amino carboxylic esters **177ec** and **177fc** were generated from the corresponding α-diazocarbonyl compounds with *p*-anisidine **165c** *via* copper-catalyzed N–H insertion with 44% and 43% yields, respectively. An oxidative deprotection of the *p*-methoxy phenyl (PMP) protecting group by means of periodic acid and subsequent work-up with hydrochloric acid in ethyl ester yielded the ammonium salts **178ec** and **178fc**. In contrast to the corresponding free amines, the ammonium salts are insensitive to oxidation and can therefore be stored at room temperature for several months. In the next step, the ammonium salt was converted to a free amine by treatment with trimethylamine and then condensed with crotonaldehyde to generate the imines **179ec** and **179fc**.[139]

The hydrolysis-sensitive imines can be transformed to the corresponding enamines *in situ* with benzyl chloroformate, which can then be converted to the desired hexahydroindole **105** *via* a diastereoselective, intramolecular DIELS-ALDER reaction. Since these procedures have already been confirmed in former experiments,[135,136,139] the same reactions have not been repeated in this work. However, the copper-catalyzed N–H insertion provides an alternative method to generate the key intermediate to synthesize the hexahydroindole motif **105**.

4.4 Chemoenzymatic Synthesis of *O*-Containing Heterocycles *via* Ketoreductase-catalyzed Highly Enantioselective Reduction

The asymmetric reduction of ketones has proven to be a fundamental strategy to produce chiral alcohols, which represents one of the most important chiral building block in the synthesis of active pharmaceutic ingredients (APIs),[120] flavors, agrochemicals and natural products. Compared to the classic synthetic methods by using sodium hydride, sodium borohydride and CBS (COREY-BAKSHI-SHIBATA) catalyst, the biocatalysis gives products with high enantio- and regioselectivity and the reaction can be performed under mild conditions. Considering the thermostability of α-diazocarbonyl compounds and the property that diazo can be decomposed by transition metal to *in situ* generate metal carbene, the enzyme-catalyzed reduction provides a practical method to retain the diazo function while the ketone is reduced to a chiral alcohol. The obtained product can on the one hand be used in intramolecular O–H insertion reactions (Scheme 39, **b**) to access *O*-containing heterocycles, on the other hand, the obtained chiral substrate can be used for further functionalization.

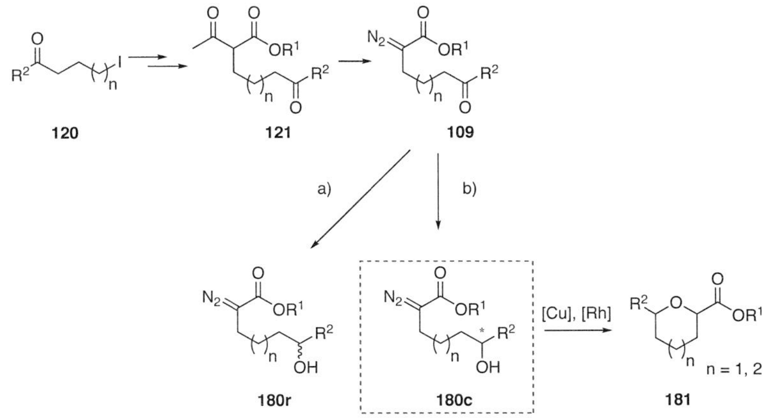

Scheme 39. Schematic synthesis of the racemic hydroxyl α-diazoesters **180r**, the chiral hydroxyl α-diazoesters **180c** as well as the *O*-containing heterocycles *via* metal-catalyzed intramolecular O–H insertion reaction. R^1 = Me, Bn. R^2 = Me, Ph. Reaction conditions: *a) NaBH4, MeOH, r.t., 30 min. b) LbADH or Gre3p.*

As already mentioned in chapter 4.1.2, the ketone-contained α-diazocarbonyl compounds can be synthesized in three steps *via* a FINKELSTEIN reaction, nucleophilic substitution followed by a classic diazo transfer reaction. The synthesized α-diazocarbonyl compounds (Figure 25) can be partly reduced by sodium borohydride to racemic hydroxyl derivatives in quantitative yields (Scheme 39, **a**), which can then be used to confirm the enantiomers generated from the enzymatic reduction as well as the metal-catalyzed O–H insertion to find out the optimal conditions to generate O-containing heterocycles.

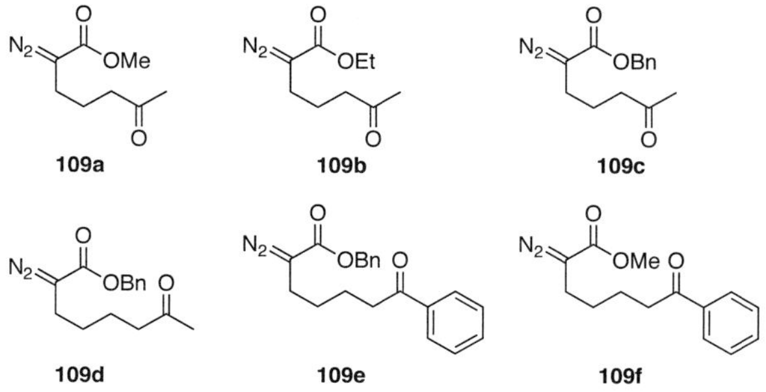

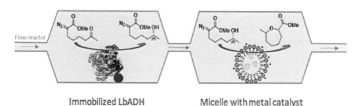

Figure 25. The synthesized α-diazocarbonyl compounds **109a–109f**.

The α-diazocarbonyl compounds can also be used in the enzymatic reduction reactions. The reactivity (turnover rate) and enantioselectivity of various ketoreductases will be screened here to find out the most efficient biocatalyst for this system. Moreover, the obtained chiral alcohols can be directly used in a consecutive reaction sequence to access the O-containing heterocycles in a flow reactor (Figure 26) with the optimized conditions obtained from the racemic substrate.[197–199] The use of a flow reactor can enable to scale up the reaction in order to generate product in large scale.

Figure 26. Illustration of a flow reactor for the consecutive synthesis of O-containing heterocycles from α-diazocarbonyl compound **109d**. The process takes place in an aqueous environment. In the first step, the chiral alcohol is generated *via* an enzyme-catalyzed enantioselective reduction. For the second step, the metal catalyst is immobilized in a micelle for the O–H insertion reaction.

4.4.1 Enzyme-catalyzed Reduction from Keto α-Diazocarbonyl Compounds

Ketoreductases (KREDs) are co-factor-dependent enzymes, which use a nicotinamide adenine dinucleotide (NADH or NADPH) as co-factor to perform the reduction. Since NADPH is too expensive for stoichiometric use and sensitive to hydrolysis, a successful regeneration of the co-factor is crucial to make the enzyme-catalyzed reduction practical. A possible system for the cofactor regeneration is the substrate-coupled regeneration with isopropanol (Figure 27).

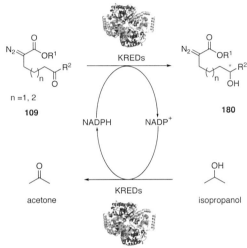

Figure 27. Substrate-coupled NADPH regeneration system. NADP⁺ is reduced to NADPH by using ketoreductase *via* oxidation of isopropanol. Isopropanol is added as a sacrificial substrate in excess to avoid the reverse reaction.

The ketoreductase catalyzes not only the reduction of the target molecular, but also the oxidation of the isopropanol. The prerequisite is that the corresponding ketoruductase accepts isopropanol as a substrate. Since it is an equilibrium reaction, a sufficiently high isopropanol concentration must be used to influence the equilibrium direction accordingly. The volatile acetone can be easily removed from the solution during the workup. Moreover, in the implementation of the poorly soluble substrate, the addition of isopropanol as a cosolvent is helpful to improve the solubility.

Alternative enzymatic cofactor regeneration systems are also available, as shown schematically in Figure 28, which uses glucose dehydrogenase (GDH) to catalyze the oxidation of D-glucose to D-Glucono-1,5-lactone. The spontaneous hydrolysis of the initial product to form gluconic acid makes the overall reaction strongly exothermic and thus favorable for the regeneration of NAD(P)H. Except the properties of readily available, inexpensive and stable, the reductant, D-Glucose is also innocuous to nicotinamide cofactors and actually increases the stability of some enzymes in solution.[115] However, the resulting gluconic acid can lead to acidification of the reaction solution, so base must be added to optimize the condition. If the corresponding substrate is base labile, an alternative regeneration system must be used.

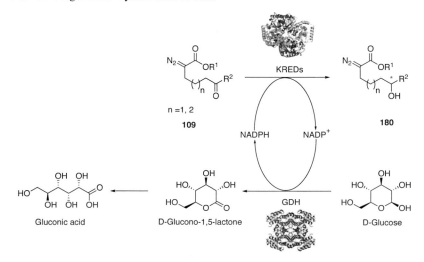

Figure 28. Enzyme-linked regeneration system. The regeneration takes place *via* the reduction of D-glucose to gluconic acid with glucose dehydrogenase (GDH) as a second enzyme.

4.4.2 Characterization of the Enzyme Activity *via* NADPH Consumption

In order to find out the optimized enzyme to reduce the ketones to corresponding chiral alcohols, various ketoreductases (KREDs), such as *Lactobacillus brevis* (LbADH), *Lactobacillus kefir* (LkADH), yeast (yGre2P, yGre3p, yPr1Pnh6, yGCY1) and the thermophilic *Alicyclobacillus acidocaldarius* (Aaci2666, Aaci2394, Aaci1696, Aaci0910) were tested and characterized.[VI] In terms of quantitation, enzymatic reduction involving NAD or NADP take advantage of the property of the reduced forms, NADH or NADPH respectively, to absorb light at a wavelength of 340 nm while the oxidized forms do not. Likewise, the reduced forms are also capable of fluorescent emission at 445 nm when excited at 340 nm, while the oxidized forms are not. These two physical properties allow investigators to quantitate reactions that directly involve a change in the oxidative state of

[VI] The research presented in this chapter has been conducted with Prof. Dr. CHRISTOPH NIEMEYER from the Institute for Biological Interfaces (IBG-1) at the Karlsruhe Institute of Technology. Esther Mittmann and Theo Peschke are thanked for their contribution to the characterization of enzyme activities.

these co-factors. The unit of the reaction rate after normalization of the enzyme concentration in the batch is defined as U/mg. This corresponds to the amount of the substrate in μmol, which is reacted by one mg of the enzyme in one minute. Taking keto α-diazocarbonyl substrate **109a** as an example, the measurement as well as the processing of the data is shown as an overview in Figure 29. Among all the ketoreductases, LbADH showed extremely high reactivity that 16.25 μmol methyl 2-diazo-6-oxoheptanoate **109a** can be transferred per minute.

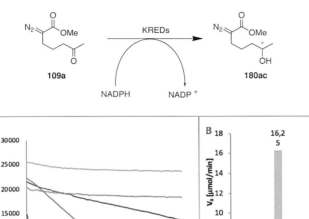

Figure 29. Enzyme activity study from the reduction of **109a** in the presence of ketoreductases. **A.** NADPH-fluorescence spectroscopy at wavelength of 445 nm. **B.** Calculated reaction rate V_0 in nmol/min by a linear adjustment in the linear region and a normalization of the enzyme concentration.

By taking advantage of the same data processing procedure, a screen of fluorescence decrease in the consumption of the cofactor NADPH was carried out to characterize the transformational activity of the different KREDs. The results of the reaction rates of different substrate-enzyme-combinations (Table 15) were determined at a substrate concentration of 1 mM. Among them, LbADH and LkADH are known in literature for their broad substrate scope.[200–203]

Table 15. Rates of turnover [μmol/(mg*min) = rpm] of ketoreductases in the reduction of various keto α-diazocarbonyl substrates **109a–109f**.

Entry	Enzyme	109a	109b	109c	109d	109e[b]	109f(32)	Expected selectivity[a]
1	LbADH	16.3	13.0	6.7	5.1	n.d.[c]	n.d.	R
2	LkADH	2.1	–[d]	-	3.5	-	-	R
3	yGre2p	0.04	-	3.1	2.0	-	0.6	S
4	yGre3p	0.1	-	0.1	-	-	n.d.	S
5	yPr1PNH6	-	-	0.3	0.5	-	n.d.	S
6	yGCY1	3.1	-	0.01	0.2	-	n.d.	S
7	Aaci2666	0.1	-	n.d.	-	-	n.d.	-
8	Aaci1696	-	-	-	n.d.	-	-	-
9	Aaci0910	-	-	n.d.	n.d.	-	n.d.	-

[a] The stereoselectivity reported here is based on literature data from the reduction of methyl ketones.[200–203] [b] In the reduction of phenyl-substituted ketones, no statement can be established about a possible stereoselectivity. [c] No enzyme activity was found. [d] The enzyme-substrate combination has not been studied so far.

The expected (R)-selectivity obtained from LbADH and LkADH showed a comparably high activity (Table 15, entry 1–2). Although the (R)-selective LbADH showed by far the highest rate of conversion in the reduction of the methyl ketone diazo compounds **109a**, **109b**, **109c** and **109d**, the phenyl-substituted ketone **109f** can only be transferred by yGre2p to give the corresponding alcohol in expected S configuration. For the insoluble diazo ketones **109c**, **109e** and **109f**, it is hard to clarify the decrease of NADPH fluorescence is due to the catalytic properties of the ketoreductases, or due to the denaturalization of the enzymes because of the insoluble oil droplets in the buffer. The purified thermophilic KREDs from *Alicyclobacillus acidocaldarius* (Aaci 2394, Aaci 1696) showed no activity in these screenings. The main reason can be attributed to the temperature: the optimum temperature for these three KREDs is 55 °C.

However, the fluorescence-based analysis only allows indirect tracking of the enzyme activity *via* degradation of the cofactor NADPH, the direct detection of products, byproducts as well as the enantioselectivities are not accessible. Thus, a chiral HPLC-based direct characterization of obtained products was carried out in the next step.

4.4.3 Characterization of the Enzyme Activity *via* Chiral HPLC

Considering the solubility as well as the ease to detect the product after intramolecular O–H insertion (an aromatic group in the product can make the product visible under the UV lamp), benzyl 2-diazo-7-oxooctanoate **109d** was chosen as a model substrate to study the enzyme activity *via* checking the enantiomer excess (ee) value of the obtained product during the reaction. The reaction was carried out by using isopropanol or glucose dehydrogenase (GDH) to *in situ* regenerate cofactor NADPH from NADP$^+$ (Figure 30, **A**). The above mentioned KREDs have been used to test the reactivity and enantioselectivity.

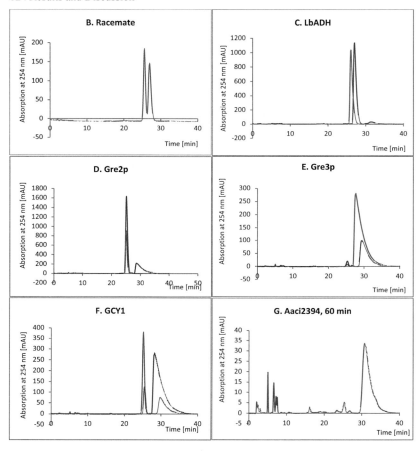

Figure 30. Comparison of the enantioselectivities of various ketoreductases *via* HPLC. **A)** Ketoreductases-catalyzed reductions from keto α-diazocarbonyl compound **109d**. **B)** HPLC spectrum of the racemate. **C)** LbADH-catalyzed reduction after 10 min and 60 min. **D)** Gre2p-catalyzed reduction after 10 min and 60 min. **E)** Gre3p-catalyzed reduction after 10 min and 60 min. **F)** GCY1-catalyzed reduction after 10 min and 60 min. **G)** Aaci2394-catalyzed reduction after 60 min. HPLC conditions (for all figures): LUX® 5 μm Amylose-1 column, n-heptane/2-propanol = 85:15, λ = 254 nm. Blue: 10 min. Red: 60 min.

It is noted LbADH showed extremely high reactivity that 94.9% ee was achieved after 10 min. Compared to LbADH, the other two expected (*R*)-selective reductases, Gre2p and GCY1 showed relatively low reactivity, but Gre2p showed better enantioselectivity than GCY1 after reaction for one hour. Gre3p performed excellent expected (*S*)-selectivity.

Of the investigated thermophilic ketoreductases Aaci1696, Aaci2394 and Aaci2666, only Aaci2394 showed low reactivity. In contrast to the fluorescence decrease tests, the reactions here are carried out at 55 °C, which may explain the different results compare to the data shown in Table 15. Aaci2394 can be assigned to (S)-selectivity.

An overview of the enzyme activity from the reduction of benzyl 2-diazo-7-oxooctanoate **109d** in the presence of ketoreductases obtained from the decrease of cofactor NADPH-fluorescence and enantiomer excess of the obtained product by means of chiral HPLC can be found in Figure 31. Due to the good selectivity of LbADH and Gre3p showed in the reaction, they are selected for use in the preparative approach to generate enantiopure products later.

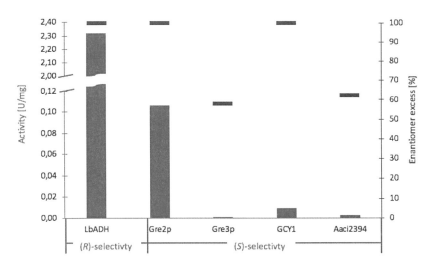

Figure 31. Enzyme activities of various ketoreductases in the reduction of substrate **109d**. The enzyme activity is determined by fluorescence decrease of NADPH (Blue, column chart) and enantiomer excess (red, scatter chart). In all approaches shown here, the cofactor regeneration system *via* glucose dehydrogenase was chosen. The unit of the reaction rate is defined as U/mg. The enantiomer excess is calculated after 60 min. HPLC conditions: LUX® 5 μm Amylose-1 column, n-heptane/2-propanol = 85:15, λ = 254 nm.

4.4.4 Transition Metal-Catalyzed Intramolecular O–H Insertion

The oxygen-containing heterocycles are ubiquitous in biologically active natural products and medicinal agents. In recent years, several synthetic methods have been developed to access oxygen-containing heterocycles. These methods are literally based on two main strategies, namely the formation of a C–C bond or C–O bond.[204]

The formation of a C–O bond has proven to be an efficient and reliable method. The most important methods used to form a C–O bond are S_N1 and S_N2 nucleophilic substitution, 1,4-conjugate addition,[205] nucleophilic ring-opening of epoxides,[206,207] metal-promoted cyclization,[208,209] hemiketalization/dehydration sequence, and hemiketalization/nucleo-philic addition sequence. To meet the growing interest in the development of eco-friendly, inexpensive and low-pollution chemical process, catalytic reactions has been studied to obtain the oxygen-containing heterocycles. Among them, the transition-metal-catalyzed insertion of α-diazo compounds into O–H bonds showed high selectivity and synthetic utility.[59,210,211]

In order to combine bio- and metal-catalysis in one process, a consecutive reaction condition should be confirmed. Considering the fact that the enantiopure alcohols reduced by the ketoreductases can only be realized in small scale, the racemic alcohols are used here as a model system to find out the optimal conditions for the metal-catalyzed intramolecular O–H insertion (Table 16).

All the reactions worked smoothly and gave products in up to 81% yield with a diastereomeric ratio of 82:18 when rhodium catalysts were used (Table 16, entries 1–7). Compared with using $Rh_2(OAc)_4$ as catalyst, $Rh_2(cap)_4$ resulted in higher yields, but lower diastereoselectivity (Table 16, entries 6–10). This is due to the electron-withdrawing nature of $Rh_2(cap)_4$, which has an influence on the reactivity *via* the resulting electrophilicity of the rhodium-carbene complex. The temperature has no significant influence on reactivity and stereoselectivity (Table 16, entries 1–2). The use of non-polar solvents resulted in better diastereoselectivity (Table 16, entries 1–3) compared to polar-aprotic solvents (Table 16, entries 4, 5, 9). The diastereomeric ration (d.r. value) showed no difference when a chiral substrate was subjected to the same conditions (Table 16, entries 1, 11), which indicates that the chiral starting material is not the key factor to influence the diastereoselectivity in the formation of C–O bond in this reaction. In contrast, the CuOTf·Tol$_{1/2}$-catalyzed O–H insertion resulted in a diastereomeric ratio of 93:7 (Table 16, entry 12), which suggests the importance of the generated metal-carbene precursor. We suppose that a higher diastereoselectivity can be realized by increasing the steric hindrance of the α-diazoesters as well as the carbene precursors.

Table 16. Metal-catalyzed intramolecular insertion of O–H bond into α-diazocarbonyl compounds (*rac*)-**180d**: reaction condition optimization.

(rac)-180d → 181d

Rh₂(OAc)₂, solvent, temperature

Entry[a]	Catalyst	Solvent	Temperature [°C]	Yield [%][b]	d.r. [%][c]
1	**Rh₂(OAc)₄**	**Toluene**	**110**	**58**	**82:18**
2	Rh₂(OAc)₄	Toluene	80	53	82:18
3	Rh₂(OAc)₄	Benzene	80	53	82:18
4	Rh₂(OAc)₄	DCE	83	50	76:24
5	Rh₂(OAc)₄	CH₂Cl₂	40	44	60:40
6	**Rh₂(cap)₄**	**Toluene**	**110**	**81**	**75:25**
7	Rh₂(cap)₄	Toluene	80	75	77:23
8	Rh₂(cap)₄	Benzene	80	55	75:25
9	Rh₂(cap)₄	CH₂Cl₂	40	76	60:40
10	[Rh(cod)Cl]₂	Toluene	110	51	75:25
11[d]	Rh₂(OAc)₄	Toluene	110	50	82:18
12[d]	CuOTf·Tol₁/₂	Toluene	110	19	93:7
13	Cu(MeCN)₄	Toluene	110	traces	_[e]

[a] Reaction conditions: catalyst (1.0 mol%), (*rac*)-**180d** (1.00 equiv.) in 2.0 mL of dry solvent was stirred under argon atmosphere for 2 hours. [b] Isolated yields. [c] The d.r. values are determined by HPLC and ¹H NMR signal integration. [d]The reaction was carried out with (*R*)-**180d** as starting material. [e] Not determined. d.r: diastereomeric ration. COD: 1,5-cyclooctadiene. Cap: tetracaprolactamate. DCE: dichloroethane.

To our delight, up to 99% enantiomer excess from both diastereomer can be obtained when the starting material was replaced by enantiopure alcohol (R)-**180d** (Table 17) and the reaction was performed under room temperature in dry toluene with the participation of $Rh_2(OAc)_4$ as catalyst. The original graphic data can be found in Chapter 6.2.9

Table 17. $Rh_2(OAc)_4$-catalyzed intramolecular insertion of O–H bond into α-diazocarbonyl compound **180d**: the effect of starting material on the enantioselectivities of products.[a]

Entry	Starting material	Temperature [°C]	d.r [%]	e.e [%] [b]	
				F1	F2
1	(rac)-180d	80	87:13	rac	rac
2	(R)-180d	80	75:25	92	92
3	**(R)-180d**	**r.t.**	**73:27**	**>99%**	**>99%**

[a] Reaction conditions: $Rh_2(OAc)_4$ (1.0 mol%) and **180d** (1.00 equiv.) in 2.0 mL of dry toluene was stirred under argon atmosphere for 2 hours. [b]HPLC condition: PHENOMENEX Amylose 2 column; n-heptane/2-propanol = 98:2; 1.5 mL/min; 30°C, 260 nm UV detector.

Theoretically, the configuration of each diastereomer can be confirmed by the difference of the spin-spin coupling in a Nuclear Overhauser effect (NOE) spectroscopy through space. But in this case, the methylene signals from the product benzyl 7-methyloxepane-2-carboxylate **181** severely overlap in the high field region. The difficulty in assigning the signals from the methylene groups makes this analysis complicated. However, the enantiopure alcohol obtained from the ketoreductase, specifically speaking, LbADH significantly improved the enantioselectivities of each diastereomer from the obtained product. We suppose with an optimization of the metal catalyst, the diastereo- and enantiopure product can be obtained.

5 Summary and Outlook

α-Diazocarbonyl compounds are highly versatile and synthetically useful reagents. In this thesis the syntheses and applications of α-diazocarbonyl compounds in transition-metal-catalyzed intermolecular N–H and intramolecular O–H insertion reactions have been described.

5.1 Synthesis of α-Diazocarbonyl Compounds

In the first part of the work, eleven saturated and six unsaturated α-diazocarbonyl compounds were generated *via* diazo transfer method (Figure 32). The use of potassium fluoride as a base allowed the synthesis of unsaturated α-diazocarbonyl compounds with ease.

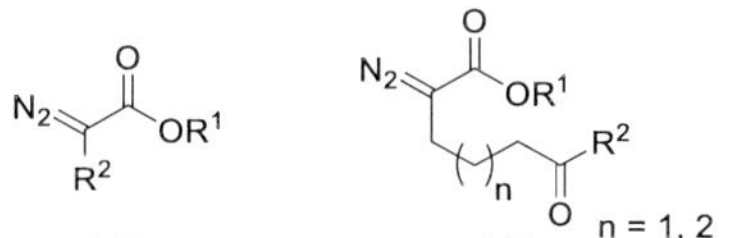

113

R^1 = Me, Et, Bn, Ph
R^2 = Me, Bn

5 examples

109

R^1 = Me, Et, Bn
R^2 = Me, Ph

6 examples

103

R^1 = Me, tBu, Bn
R^2 = Me, Et, tBu, Bn

6 examples

Figure 32. The synthesized α-diazocarbonyl compounds.

Outlook

However, the total yields after two to four reaction steps are not satisfying. Especially the nucleophilic addition step gives low yields and the reaction scope is limited to alkyl and aryl moieties. Thus, a new method to generate substituted β-keto esters is necessary.

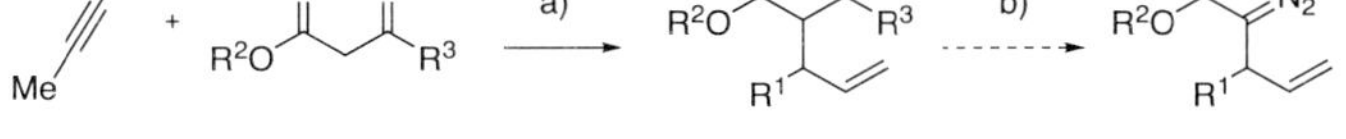

182   **183**   **184**   **185**

Scheme 40. Rhodium-catalyzed regioselective addition of β-keto esters **183** to internal alkynes **182**. Reaction conditions: a) *[Rh(cod)Cl]$_2$ (2.0 mol%), DPEphos (6.0 mol%), TFA (20 mol%), DCE/EtOH (5:1), 80 °C, 16 h. R^1 = alkyl, aryl. R^2 = alkyl, Ph. R^3 = alkene, aryl, CH$_2$Cl$_2$, CF$_3$, heterocycle. b) p-ABSA (2.00 equiv.), Et$_3$N (3.00 equiv.), CH$_3$CN, r.t.*

A regioselective rhodium-catalyzed addition of β-keto esters to internal alkynes (Scheme 40) can be carried out to enlarge the substrate scope and improve the reaction yield.[129] In addition, when the diazo transfer method is also effective for this kind of β-keto esters and $R^1 \neq$ aryl, $R^2 \neq$ aryl, the water solubility of the obtained α-diazocarbonyl compounds would be improved. This could provide a lot of useful substrates for enzyme-catalyzed reductions.

5.2 [2.2]Paracyclophane-based Catalysts and Metal-catalyzed N–H Insertion Reactions

The second part of the thesis focused on the synthesis of catalysts with a [2.2]paracyclophane backbone. Three novel enantiopure (R_p)-pseudo-*ortho* [2.2]paracyclophane-based bisoxazoline ligands (R_p)-**144a–144c** were obtained from unsubstituted [2.2]paracyclophane *via* bromination, microwave-assisted isomerization, HPLC-based chiral separation, lithiation-carboxylation and condensation (Scheme 41).

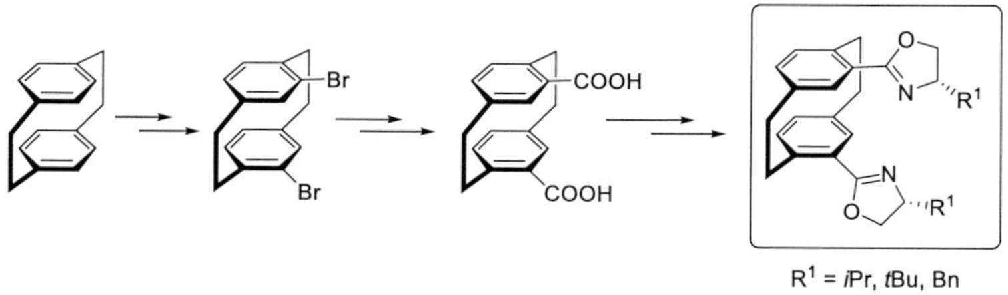

$R^1 = i$Pr, tBu, Bn

| **59** | (R_p)-**75** | (R_p)-**146** | (R_p)-**144a-144c** |

Scheme 41. Schematic synthesis of (R_p) pseudo-*ortho* [2.2]paracyclophane-based bisoxazoline ligands (R_p)-**144a-144c**.

The ruthenium-catalyst with the [2.2]paracyclophane backbone **59** was obtained *via* cycloruthenation from mono-phenyloxazoline-substituted [2.2]paracyclophane $(S_p,S)/(R_p,S)$-**155** and [RuCl$_2$(p-cymene)]$_2$ in 54% yield (Scheme 42). As a modification of Ru(II)-Pheox **156**, the acetonitrile-coordinated ruthenium complex **157** was obtained *via* cycloruthenation from $(S_p,S)/(R_p,S)$-**155** and [RuCl$_2$Ph]$_2$ (Scheme 42). The complex showed high instability and getting decomposed during the ^{13}C NMR measurement. However, the *in situ* generated **157** yielded 55% product in the insertion of unsaturated α-diazocarbonyl **103e** into the N–H bond of aniline **165a**.

Compared to ruthenium-catalysts **151** and **157**, the *in situ* generated copper catalysts with pseudo-*ortho* [2.2]paracyclophane-based bisoxazoline ligands (R_p,S)-**144c/144b** performed excellent reactivity in the insertion of **103e** into the N–H bond of aniline **165a** ((R_p,S)-**144c**, 88%; (R_p,S)-**144b**, 80%).

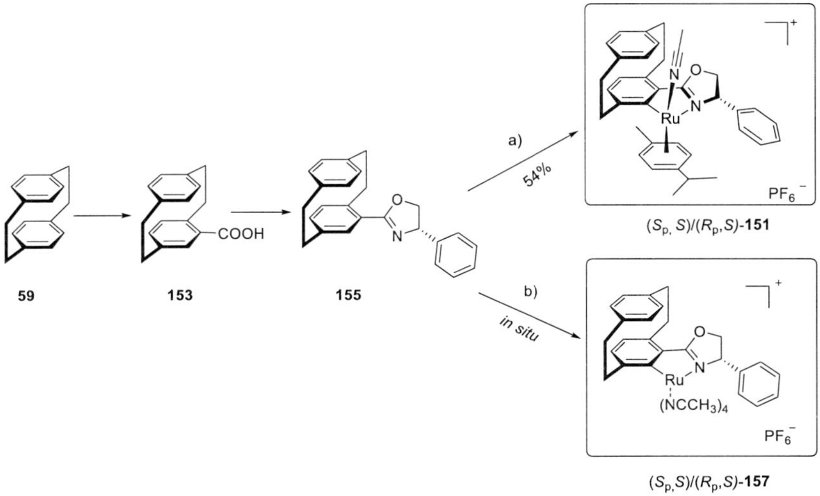

Scheme 42. Schematic synthesis of the Ru(II)-[2.2]paracyclophane complex from 4-phenyloxazoline-substituted [2.2]paracyclophane ligand **155**. Reaction conditions: *a) [RuCl₂(p-cymene)]₂ (0.50 equiv.), KOAc (1.50 equiv.), KPF₆ (2.00 equiv.), CH₃CN, r.t., 3 d; b) [RuCl₂Ph]₂ (0.50 equiv.), KOAc (1.50 equiv.), KPF₆ (2.00 equiv.), CH₃CN, 80 °C, 80 h.*

We propose that the lack of enantioselectivity observed from the Cu-catalyzed insertion of α-diazocarbonyl compound **103e** into the N–H bond of aniline **165a** in the presence of (R_p)-pseudo-*ortho* [2.2]paracyclophane-based bisoxazoline ligands **144a–144b** occurs due to the inherent conformational rigidity of the ligand backbone and the distance between the two oxazoline groups in the molecule. The crystallographic data shows that the distance between the two nitrogen atoms is 5.330 Å and 5.089 Å, which means that the copper carbene (Figure 22) cannot be fixed in the 'pocket' generated by the bisoxazoline ligand. Therefore, the lone pair electrons of the aniline N atom can attack the carbene without selectivity to generate the ylide **174**.

With the employment of the *in situ* generated Cu-(S,S)-Ph-Pybox **172** catalyst, twenty-one different δ-amino α,β-unsaturated carboxylic esters **177** were obtained in 14–75% yields (Scheme 43). The study on the influence of the anilines demonstrated that strong electron withdrawing or donating groups on the aniline lead to a decrease in the corresponding yields. The amino-substituted [2.2]paracyclophanes **161** and **161a** are compatible with this catalytic system, which provides an alternative method to modify amino-substituted [2.2]paracyclophane.

The obtained δ-amino α,β-unsaturated carboxylic esters **177ec** and **177fc** underwent oxidative deprotection followed by condensation with crotonaldehyde to generate the imine intermediates **179ec** and **179fc**. These imines can then be *in situ* transformed to enamines with benzyl chloroformate, which are the diene-dienophile precursors for the intramolecular DIELS-ALDER reaction to synthesize the hexahyroinoles **105**.

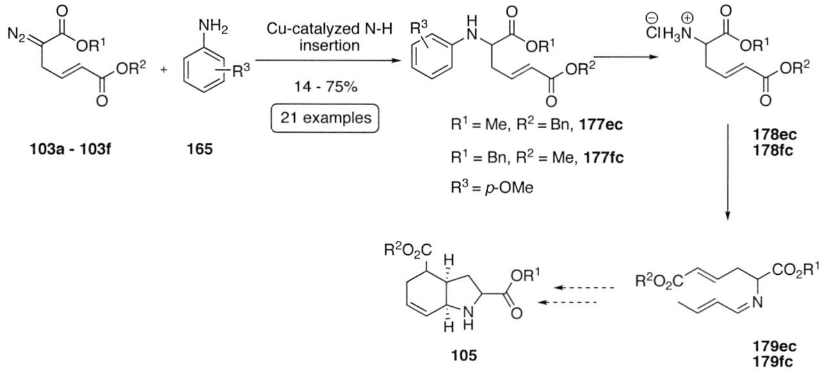

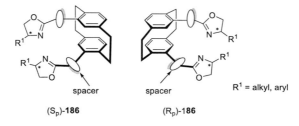

Scheme 43. Synthesis of the hexahydroindole motif **105** *via* a copper-catalyzed N–H insertion from unsaturated α-diazocarbonyl compounds.

Outlook

Considering the inherent conformational rigidity provided by the [2.2]paracyclophane backbone, the distance between the N–N bond from the bisoxazoline cannot be easily changed by introduction of functional groups to the oxazoline core. Thus, a spacer could be introduced to the pseudo-*ortho* position of the [2.2]paracyclophane backbone to offer the conformational flexibility, which makes to distance between the two oxazoline groups suitable for performing (Figure 33).[59] Moreover, the steric or electronic element based on the spacer could also affect the electron density on the N atoms from bisoxazoline and so influence the ability to coordinate with a metal.

Figure 33. Modification of the pseudo-*ortho* [2.2]paracyclophane-based bisoxazoline ligands by introduction of a spacer to improve the conformational flexibility.

5.3 Chemoenzymatic Synthesis of *O*-Containing Heterocycles *via* Ketoreductase-catalyzed Highly Enantioselective Reduction

In this work, various ketoreductases from different mesophilic and thermophilic bacteria and yeasts have been used as effective chiral biocatalysts in the enantioselective reduction of α-diazocarbonyl compounds. Six substrates **109a–109f**, bearing different side chain lengths and keto substitutions have been studied (Figure 34).

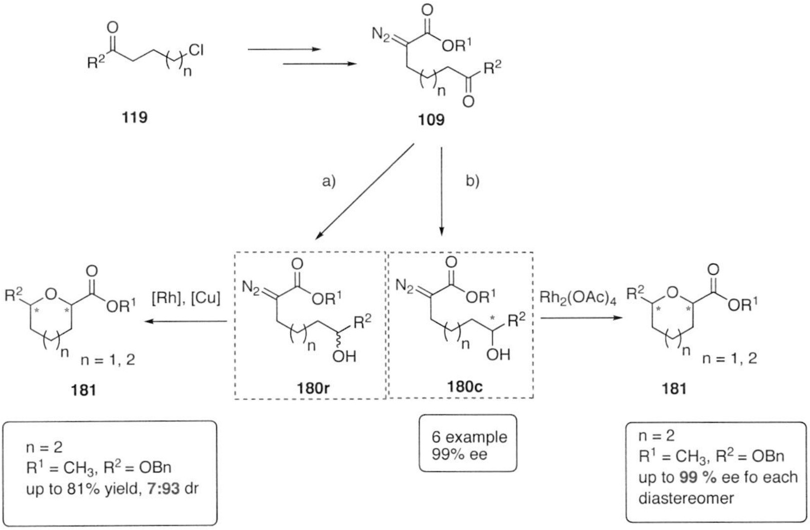

Figure 34. Chemoenzymatic synthesis of the *O*-containing heterocycles. **a)** Reduction *via* NaBH₄. **b)** Enzyme-catalyzed highly enantioselective reduction.

The highest turnover rates have been achieved with the (*R*)-selective LbADH and the (*S*)-selective Gre3p. Both of them showed high reactivity (≥ 0.2 μmol/(min*mg)) and excellent enantioselectivity (≥ 99% ee) (Figure 31).

The intramolecular insertion of O–H bond into the racemic benzyl-2-diazo-7-hydroxyheptanoate (*rac*)-**180d** yielded up to 81% of the product **181d** when Rh₂(OAc)₄ was used as a catalyst. CuOTf·Tol₁/₂ resulted in 7:93 diastereoselectivity. Up to 99% enantiomer excess from both diastereomer can be obtained when the starting material was replaced by the enantiopure alcohol (*R*)-**180d** and the reaction was performed under room temperature in dry toluene with the participation of Rh₂(OAc)₄ as catalyst (Figure 35).

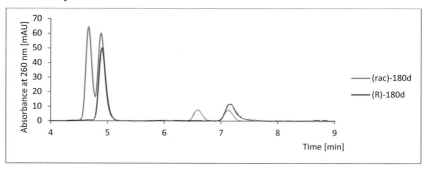

Figure 35. Chiral HPLC spectra of the obtained *O*-containing heterocycles **181d**. Green: (*rac*)-**180d** as starting material, both diastereomers showed no enantioselectivity. Red: (*R*)-**180d** as starting material, up to 99% e.e from both diastereomer can be obtained. HPLC condition: PHENOMENEX Amylose 2 column; n-heptane/2-propanol = 98:2; 1.5 mL/min; 30°C, 260 nm UV detector.

Outlook

Since the solubility of the substrates in aqueous media has a big influence on the enzyme activity, α-diazocarbonyl compounds with hydrophilic side chains should be synthesized. By using the method depicted in Scheme 40, the substituted β-keto esters with alkene side chains can be generated *via* rhodium-catalyzed addition of β-keto esters to internal alkynes. If these substrates are also compatible for the diazo transfer method, the solubility of the obtained α-diazocarbonyl compounds could be improved.

Although the enantioselectivity of each diastereomer have been effectively improved by the employment of enantiopure starting material, the diastereoselectivity have not been really optimized. The dirhoduium (II) paddlewheel complexes developed by DAVIES *et al.* showed remarkable performance in the carbenoid chemistry.[212–214] Instead of the acetate group on $Rh_2(OAc)_4$, a substitution with electronic and/or steric properties could change the nature of the ligands, which therefore influence the property of the *in situ* generated rhodium-carbene. The well-known proline derived rhodium (II) carboxylates,[215,216] phthalimide derived rhodium (II) carboxylates[217–221] as well as a Rh(II)-paracyclophane catalyst can be applied in the future work to optimize the diastereoselectivity of the products.

The combination of enzyme-catalyzed reduction and the optimized O–H insertion in a flow reactor could enable a large-scale synthesis of *O*-containing heterocycles with optimal diastereo- and enantioseletivities. The active site of the enzyme can be accessed by means of enzyme immobilization. The hydrophobic metal catalyst can be immobilized in the micelle system to improve its availability (Figure 26).

6 Experimental Part

6.1 General Information on [2.2]Paracyclophane

6.1.1 Nomenclature of [2.2]Paracyclophane

The IPUAC nomenclature for cyclophanes is not intuitive and difficult to understand.[222] Therefore a specific cyclophane nomenclature has been developed by VÖGTLE *et al.*, which is based on the core-substituent ranking.[223] The core structure is named according to the length of the bridges (number of atoms between two aromatic rings) and put in front of the name in squared brackets (e.g. [m,n]) and the suffix "phane" is added. Following this the substitution patterns on the benzene ring signed by *ortho*, *meta* or *para*. An example is shown in Figure 36.

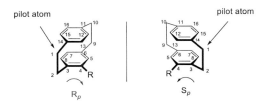

Figure 36. Entire nomenclature shown on mono-substituted [2.2]paracyclophane.

[2.2]Paracylophane belongs to the D_{2h} point group, which is broken by the first substituent, resulting in two planar chiral enantiomers. For the determination of their absolute configuration, the arene bearing the substituent is set as the chiral phane, and the first atom outside the phane and closest to the chirality center is set as the "*pilot atom*". If both arenes are substituted, then the substituent with higher priority according to the CAHN-INGOLD-PRELOG (CIP) nomenclature is preferred.[224] The stereo descriptor is determined by the sense of rotation viewed from the pilot atom. An unambiguous numeration (Figure 36) is required to describe the positions of the substituents correctly.

6.1.2 Analytics and Equipment

Nuclear Magnetic Resonance Spectroscopy (NMR):

NMR spectra have been recorded using the following machines:

^{1}H NMR: BRUKER *Avance* 300 (300 MHz), BRUKER *Avance* 400 (400 MHz), BRUKER *Avance DRX 500* (500 MHz). The chemical shift δ is expressed in parts per million (ppm) where the residual signal of the solvent has been used as secondary reference: chloroform (^{1}H: δ = 7.26 ppm), acetone (δ = 2.05 ppm), dimethyl sulfoxide (δ = 2.50 ppm) and dichloromethane (δ = 5.32 ppm). The spectra were analyzed according to first order.

^{13}C NMR: Bruker *Avance* 300 (75 MHz), Bruker *Avance* 400 (101 MHz), Bruker *Avance DRX 500* (126 MHz). The chemical shift δ is expressed in parts per million (ppm), where the residual signal of the solvent has been used as secondary reference: chloroform (δ = 77.0 ppm), acetone (δ = 30.8 ppm), dimethyl sulfoxide (δ = 39.4 ppm) and dichloromethane (δ = 53.8 ppm). The spectra were ^{1}H-decoupled and characterization of the ^{13}C NMR-spectra ensued through the DEPT-technique (DEPT = Distortionless Enhancement by Polarization Transfer) and is stated as follows: DEPT: "+" = primary or tertiary carbon atoms (positive DEPT-signal), "–" = secondary carbon atoms (negative DEPT-signal), C_q = quaternary carbon atoms (no DEPT-signal).

All spectra were obtained at room temperature. As solvents, products obtained from EURISOTOP and SIGMA ALDRICH were used: chloroform-d_1, acetone-d_6, dimethylsulfoxide-d_6 and dichloromethane-d_2, acetonitrile-d_3. For central symmetrical signals the midpoint is given, for multiplets the range of the signal region is given. The multiplicities of the signals were abbreviated as follows: s = singlet, d = doublet, t = triplet, q = quartet, hept = heptet, brs = broad singlet, m = multiplet and combinations thereof. All coupling constants J are stated as modulus in Hertz [Hz].

Infrared spectroscopy (IR):

IR-spectra were recorded on a BRUKER *Alpha T*. Measurement of the samples was conducted *via* attenuated total reflection (ATR). The intensity of bands (strength of absorption) was described as follows: vs = very strong (0–9% transmission, T), s = strong (10–39% T), m = middle (40–69% T), w = weak (70–89%), vw = very weak (90–100%). Position of the absorption bands is given as wavenumber $\widetilde{\nu}$ with the unit [cm^{-1}].

Mass Spectrometry (EI-MS, FAB-MS):

Mass spectra were recorded on a FINNIGAN *MAT 95*. Ionization was achieved through either EI (*Electron Ionization*) or FAB (*Fast Atom Bombardment*). Notation of molecular fragments is given as mass to charge ratio (*m/z*), the intensities of the signals are noted in percent relative to the base signal (100%). As abbreviation for the ionized molecule $[M]^+$ was used. Characteristic fragmentation peaks are given as $[M–fragment]^+$ and $[fragment]^+$.

For HRMS (High Resolution Mass Spectrometry) following abbreviations were used: calc. = expected value (calculated), found = value found in analysis.

Gas Chromatography-Mass Spectrometry (GC-MS):

GCMS measurements have been recorded with an AGILENT TECHNOLOGIES model *6890N* (electron impact ionization), equipped with a AGILENT *19091S-433* column (5% phenyl methyl siloxane, 30 m, 0.25 µm) and a *5975B VL MSD* detector with turbo pump. As a carrier gas helium was used.

Elemental Analysis (EA):

Measurements were conducted on an ELEMENTAR *Vario Micro*. As analytical scale SARTORIUS *M2P* was used. Notation of Carbon (C), Hydrogen (H) and Nitrogen (N) is given in mass percent. Following abbreviations were used: calc. = expected value (calculated), found = value found in analysis.

Thin Layer Chromatography (TLC):

All reactions were monitored by thin layer chromatography (TLC). The TLC plates were purchased from MERCK (silica gel 60 on aluminum plate, fluorescence indicator F_{254}, 0.25 mm layer thickness). Detection was carried out under UV-light at $\lambda = 254$ nm. Alternatively, the TLC plates were stained with a SEEBACH-stain solution (2.5% phosphor molybdic acid, 1.0% cer(IV)sulfate-tetrahydrate, 6.0% conc. sulfuric acid, 90.5% water, as dip solution) and dried in a hot air stream.

Analytical Scales:

Used device: SARTORIUS Basic.

Polarimeter:

The rotations were measured with a PERKIN ELMER 241 polarimeter at 20 °C with a glass cuvette (length = 1 dm) and the sodium D-line.

The term for the specification is given by

$$[\alpha]_D^{20} = \frac{\alpha}{\beta \times d}$$

where D is the wavelength of the light employed (sodium D-line, $\lambda = 589.3$ nm), α is the observed rotation, d is the path length (d =1), β is the concetration in grams per milliliter.

The specification of the rotation value was given like: $[\alpha]_D^{20} = -64.8$ (c = 0.58, $CHCl_3$), c = concentration.

Microwave:

Reactions heated using microwave irradiation were carried out in a single mode *CEM* Discover LabMate microwave operated with *CEM*'s Synergy software. This instrument works with a constantly focused power source (0–300W). Irradiation can be adjusted *via* power- or temperature control. The temperature was monitored with an infrared sensor.

Solvents and Reagents:

Solvents of technical quality have been purified by distillation prior to use. Solvents of the grade p.a. (*per analysis*) have been purchased (ACROS, FISHER SCIENTIFIC, SIGMA ALDRICH, ROTH, RIEDEL–DE HAËN) and used without further purification. Absolute solvents have been dried using the methods listed in Table 18, and were stored under argon afterwards or have been purchased from a commercial supplier, including abs. acetonitrile (Acros, < 0.005% water), abs. chloroform (Fischer, over molecular sieves), abs methanol (Fischer, < 0.005% water), abs. ethanol (Acros, < 0.005% water).

Reagents have been purchased from commercial suppliers (Companies: ABCR, ACROS, ALFA AESAR, APOLLO, CARBOLUTION, CHEMPUR, FLUKA, IRIS, MAYBRIDGE, MERCK, RIEDEL DE HAËN, TCI, THERMO FISHER SCIENTIFIC, SIGMA ALDRICH). They have been used without further purification unless stated otherwise.

Table 18: Methods for solvents purification. All distillations were carried out under argon atomosphere.

Solvent	Method
Dichloromethane	heating to reflux over calcium hydride, distilled over a packed column
Tetrahydrofuran	heating to reflux over sodium metal (benzophenone as an indicator), distilled over a packed column
Diethylether	heating to reflux over sodium metal (benzophenone as an indicator), distilled over a packed column
Toluene	heating to reflux over sodium metal (benzophenone as an indicator), distilled over a packed column

6.1.3 Preparative work

Before the reactions with air or moisture sensitive reagents were carried out, the glass devices have been dried in an oven and under high vacuum. After cooling the glass apparatuses have been sealed with rubber seals under argon counterflow. Reactions have been executed according to SCHLENK-techniques using argon as an inert gas. Liquids were added *via* plastic syringes and V2A-needles. Solids were added in pulverized form. Reactions at 0 °C were cooled with a mixture of ice/water. Reactions at lower temperatures were tempered with brine/ice mixture (–20 °C), isopropanol/dry ice mixture (–78 °C) or with a cryostat.

All reactions were monitored by thin layer chromatography (TLC) or GCMS.

Solvents were removed at 40 °C with a rotary evaporator. Used solvent mixtures were measured volumetrically. An ultraviolet lamp as well as the phosphomolybdic acid 10 wt% in ethanol were used for detection.

If not stated otherwise, solutions of inorganic salts were saturated aqueous solutions.

If not otherwise specified, the raw products, were purified by flash column chromatography following the concepts of Still *et al.* using silica gel (SIGMA ALDRICH, pore size 60 Å, particle size 40 – 63 μm) and sand (calcined and purified with hydrochloric acid) as stationary phase. The use of a gradient is indicated in the experimental procedures.

Celite® for filtrations was purchased from ALFA AESAR (Celite® 545, treated with Na_2CO_3).

6.2 Synthesis and Characterization

6.2.1 General Procedures

General Procedure 1 (GP 1): Synthesis of substituted buta-2,3-dienoate 143

To a solution of substituted triphenylphosphonium bromide (1.00 equiv.) in CH_2Cl_2 was added 10% NaOH aq. solution (100 mL, w/v). After 15 min of vigorous stirring, the phases were separated. The aqueous phase was extracted three times with CH_2Cl_2 and the organic phases were combined, washed with brine, dried over Na_2SO_4, filtrated and evaporated under reduced pressure to give the substituted 2-(triphenylphosphoranylidene) acetate **142** as white solid.

To a solution of **142** (1.00 equiv.) in anhydrous CH_2Cl_2 and pentane was added dry triethylamine (1.00 equiv.) under argon atmosphere. Acetyl chloride (1.10 equiv.) was added dropwise *via* a dropping funnel over a period of 1 h at room temperature. After stirring overnight, the heterogeneous mixture was filtered, the precipitate was washed with pentane and the filtrate was concentrated at 20 °C on a rotary evaporator. Pentane was added to the residue and the mixture was stirred vigorously for 30 min, the resulting white precipitate was removed by filtration and washed with pentane. The filtrate was concentrated on a rotary evaporator and the residue was purified *via* column chromatography (Pentane/Et_2O = 3:1) to give substituted buta-2,3-dienoate **143** as colorless liquids.

General Procedure 2 (GP 2): Phosphine-catalyzed umpolung addition

A two neck round-bottom flask equipped with a stirring bar and a condenser was flame-dried and left to cool under argon atmosphere. After triphenylphosphine (20 mol%) was added to a solution of β-acetoacetate (**135**, 1.10 equiv.) in distilled benzene, buta-2,3-dienoate (**143**, 1.00 equiv.) in distilled benzene was added dropwise to the reaction mixture *via* a syringe pump over 5 h. The reaction was left to proceed until **143** was consumed. The crude reaction mixture was concentrated and loaded onto a silica gel column and separated chromatographically (*c*-Hex/EtOAc = 5:1). In all cases, the obtained product **130** were stained brightly with SEEBACH-stain solution after heating.

General Procedure 3 (GP 3): Diazo transfer reaction in the synthsis of unsaturated α-diazopropionates 103

To a solution of **130** (1.00 equiv.) in dry CH_3CN was added potassium fluoride (3.00 equiv.) in portions, a solution of *p*-ABSA (2.00 equiv.) in dry CH_3CN was added dropwise to the reaction. After stirring at room temperature for 12 h, the reaction mixture was filtered, the filtrate was concentrated under reduced pressure, the resulting mixture was diluted in Et_2O

and washed with 5% aq. KOH (w/v) solution. The water phase was extracted with Et_2O three times, the combined organic layers were washed with brine, dried over Na_2SO_4, filtrated and concentrated under vacuum. The crude product was purified *via* column chromatography (*c*-Hex/EtOAc = 6:1) to give the unsaturated α-diazopropionates **103** as yellow liquid.

General Procedure 4 (GP 4): Metal-catalyzed N–H insertion reaction from the unsaturated α-diazopropionates

Method A (GP 4a): Cu(MeCN)$_4$PF$_6$ (10 mol%), ligand (12 mol%) and NaBArF (12 mol%) were added into an oven-dried screw vial, evacuated and backfilled with argon three times. After CH_2Cl_2 (1 mL) was injected into the vial, the solution was stirred at 40 °C under argon atmosphere for 2 h. A solution of unsaturated α-diazopropionates (**103**, 1.20 equiv.) in 1 mL of CH_2Cl_2 and aniline (**165**, 1.00 equiv.) were added subsequently, the mixture was stirred at room temperature for 2 h. The resulting mixture was dried under vacuum and purified directly *via* column chromatography (*c*-Hex/EtOAc = 8:1 or pentane/Et$_2$O = 5:1) to give the insertion product **177**.

Method B (GP 4b): To an oven-dried screw vial was added Ru-complex (12 mol%) and NaBArF (12 mol%) in 1 mL of abs. CH_2Cl_2. A solution of unsaturated α-diazopropionates (**103**, 1.20 equiv.) in 1 mL of CH_2Cl_2 and aniline (**165**, 1.00 equiv.) were added subsequently under argon atmosphere. After stirring at room temperature for 2 h, the resulting mixture was dried under vacuum and purified directly *via* column chromatography (*c*-Hex/EtOAc = 8:1 or pentane/Et$_2$O = 5:1) to give the insertion product **177**.

Method C (GP 4c): To an oven-dried screw vial was added $(S_p,S)/(R_p,S)$-4-(4'-phenyloxazolin-2'-yl)[2.2]paracyclophane $((S_p,S)/(R_p,S)$-**155**, 1.00 equiv.), [RuCl$_2$Ph]$_2$ (0.50 equiv.), potassium acetate (1.50 equiv.) and potassium hexafluorophosphate (2.00 equiv.) in 5 mL of abs. CH_3CN, the mixuture was stirred at 80 °C under argon atmosphere for 80 hours. After the resulting mixture was cooled to room temperature, the solvent was removed under high vacuum to give the Ru-complex **157** in bright yellow color.

The *in situ* generated Ru-complex **157** was dissolved in 1 mL of degassed CH_2Cl_2, a solution of unsaturated α-diazopropionates (**103**, 1.20 equiv.) in 1 mL of CH_2Cl_2 and aniline (**165**, 1.00 equiv.) were added subsquently, the mixture was stirred at room temperature for 2 h. The resulting mixture was dried under vacuum and purified directly *via* column chromatography (*c*-Hex/EtOAc = 8:1 or pentane/Et$_2$O = 5:1) to give the insertion product **177**.

General Procedure 5 (GP 5): Synthesis of iodo-substituted ketone 120 (FINKELSTEIN reaction)

To a solution of chloro-substituted ketone (**119**, 1.00 equiv.) in dry acetone was added sodium iodide (2.50 equiv.). After stirring under reflux overnight, the solvent was removed under vacuum, the residue was dissolved in water and extracted with CH_2Cl_2 three times. The combined organic layer was washed with brine, dried over Na_2SO_4, filtrated and concentrated under vacuum. Products **120** were obtained as yellow to dark brown liquids. The yields were determined by NMR integration according to the ration between $-CH_2Cl$ and $-CH_2I$.

General Procedure 6 (GP 6): Synthesis of keto-substituted β-acetoacetate 121

To a suspension of NaH (1.20 equiv.) in THF was added β-acetoacetate (**111**, 1.20 equiv.) in dry THF under argon atmosphere at 0 °C. After the solution turned clear, iodo-substitued ketone (**120**, 1.00 equiv.) in dry THF was added dropwise. The solution was allowed to warm to room temperature before being heated to reflux. After stirring under reflux overnight, the reaction mixture was quenched with water and extracted with EtOAc three times, the combined organic layer was washed twice with brine, dried over anhydrous Na_2SO_4, and concentrated in *vacuo*. The residue was purified *via* column chromatography to give the keto-substituted β-acetoacetate **121** as colorless liquid.

General Procedure 7 (GP 7): Diazo transfer reaction to generate α-diazopropionates 109

To a solution of **121** (1.00 equiv.) in dry CH_3CN was added Et_3N (3.00 equiv.) at 0 °C. A solution of *p*-ABSA (2.00 equiv.) in CH_3CN was added dropwise into the mixture. After stirring at room temperature overnight, the reaction mixture was concentrated under reduced pressure, washed with 5% aq. KOH and extracted with EtOAc three times, the combined organic layer was washed with brine, dried over Na_2SO_4 and evaporated under vacuum. The residue was purified *via* column chromatography to give the title product **109** as yellow liquid.

General Procedure 8 (GP 8): Syntheis of hydroxyl-subsituted α-diazocarbonyl compounds

To a solution of **109** (1.00 equiv.) in methanol was added $NaBH_4$ (2.50 equiv.) at 0 °C in portions. After stirring at 0 °C for 30 min, the solvent was removed under reduced pressure, the residure was dissolved in water and extracted with EtOAc (2 × 20 mL), the combined organic layer was washed with brine, dried over Na_2SO_4, filtrated, and evaporated under vacuum. The residue was purified *via* column chromatography to give the title product **180r** as yellow liquid.

General Procedure 9 (GP 9): Metal-catalyzed O–H insertion to generate *O*-containing heterocycles 181

To a suspension of [Rh]/[Cu] (0.002 mmol, 1.0 mol%) in abs. solvent (0.50 mL) was added a solution of benzyl 2-diazo-7-hydroxyoctanoate (**180d**, 55.2 mg, 0.20 mmol, 1.00 equiv.) in 0.50 mL of abs. solvent dropwise under argon atomosphere. After stirring under different conditions, the solvent was removed under reduced pressure and the resulting mixture was purified *via* column chromatography (*c*-Hex/EtOAc = 4:1) to give the title product **181** as colorless liquid.

6.2.2 Synthesis and Characterization of Paracyclophane Derivatives

(*rac*)-4-Bromo[2.2]paracyclophane (152)

A solution of Br_2 (3.77 mL, 11.8 g, 73.7 mmol, 1.02 equiv.) in CCl_4 (80 mL) was prepared. A suspension of iron powder (0.80 g, 3.84 mmol, 0.05 equiv.) in 5 mL of the Br_2/CCl_4 solution was stirred at room temperature for 1 h. After CH_2Cl_2 (250 mL) was added to dilute the suspension, [2.2]paracyclophane (**59**, 15.0 g, 72.0 mmol, 1.00 equiv.) was added and the mixture was stirred for 30 min. The remaining Br_2/CCl_4 solution was added dropwise over 5 h. After stirring overnight, a solution of 100 mL sat. $Na_2S_2O_3$ was added, the water phase was extracted with CH_2Cl_2 (2 × 50 mL), the combined organic phase was washed with brine, filtrated, dried over Na_2SO_4 and concentrated under vacuum. The product **152** was obtained as light yellow solid, 20.1 g, 70.3 mmol, 98%.

R_f = 0.72 (Pentane/EtOAc = 9:1). – **^{1}H NMR** (400 MHz, CDCl$_3$) δ/ppm = 7.17 (dd, J = 7.9, 2.0 Hz, 1H, CH^{Ar}), 6.56 (td, J = 8.1, 1.9 Hz, 1H, CH^{Ar}), 6.52 (s, 2H, 2 × CH^{Ar}), 6.51–6.43 (m, 3H, CH^{Ar}), 3.47 (ddd, J = 13.2, 10.0, 2.2 Hz, 1H, CH^{PC}), 3.21 (ddd, J = 13.1, 10.0, 6.0 Hz, 1H, CH^{PC}), 3.17–3.01 (m, 4H, CH^{PC}), 2.99–2.87 (m, 1H, CH^{PC}), 3.21 (ddd, J = 13.3, 10.5, 6.1 Hz, 1H, CH^{PC}). – **^{13}C NMR** (101 MHz, CDCl$_3$) δ/ppm = 141.61 (C$_q$, C^{Ar}), 139.32 (C$_q$, C^{Ar}), 139.10 (C$_q$, 2 × C^{Ar}), 137.27 (+, CH, C^{Ar}), 135.04 (+, CH, C^{Ar}), 133.30 (+, CH, C^{Ar}), 132.91 (+, CH, C^{Ar}), 132.22 (+, CH, C^{Ar}), 131.45 (+, CH, C^{Ar}), 128.70 (+, CH, C^{Ar}), 126.96 (C$_q$, C_{Ar}–Br), 35.81 (–, CH_2), 35.49 (–, CH_2), 34.82 (–, CH_2), 33.48 (–, CH_2). – **IR** (ATR): $\tilde{\nu}$/cm^{-1} = 2924 (w), 2848 (vw), 1891 (vw), 1585 (vw), 1541 (vw), 1496 (vw), 1475 (vw), 1431 (vw), 1408 (vw), 1390 (w), 1185 (vw), 1091 (vw), 1034 (vw), 940 (vw), 896 (w), 839 (w), 793 (w), 708 (w), 667 (w), 640 (w), 576 (w), 513 (w), 473 (w), 403 (vw), 381 (vw). – **MS** (EI, 70 eV), m/z (%): 288/286 (100/96) [M]$^+$, 184/182 (15/15) [M – C$_8$H$_8$]$^+$, 104 (18) [C$_8$H$_8$]. – **HRMS** (EI, C$_{16}$H$_{15}$^{79}Br$_1$) calc. 286.0352, found 286.0352. The analytical data matched those reported in the literature. [75,225–227]

4,16-Dibromo[2.2]paracyclophane (71)

A solution of Br_2 (5.50 mL, 17.0 g, 106 mmol, 2.20 equiv.) in CH_2Cl_2 (50 mL) was prepared. A suspension of iron powder (0.14 g, 2.4 mmol, 0.05 equiv.) in 6.25 mL of the Br_2/CCl_4 solution was diluted in 50 mL of CH_2Cl_2 and stirred at room temperature for 1 h. The solution was then brought to reflux for 2 h. CH_2Cl_2 (50 mL) and [2.2]paracyclophane (**59**, 10.0 g, 48.0 mmol, 1.00 equiv.) were added to the mixture subsequently. After the remaining bromine solution was added dropwise over a period of 4 h, the mixture was stirred at room temperature for 3 days. Saturated $Na_2S_2O_3$ solution was added and the reaction mixture was stirred at room temperature until the bromine color disappeared. The organic phase was separated and filtrated, the precipitate was recrystallized from hot toluene to obtain product **71** as off-white solid, 5.40 g, 14.8 mmol, 31%.

R_f = 0.75 (*c*-Hex/EtOAc = 90:10). – **¹H NMR** (400 MHz, CDCl₃) δ/ppm = 7.14 (dd, J = 7.8, 1.8 Hz, 2H, 2 × CH^{Ar}), 6.51 (d, J = 1.8 Hz, 2H, 2 × CH^{Ar}), 6.44 (d, J = 7.8 Hz, 2H, 2 × CH^{Ar}), 3.50 (ddd, J = 12.8, 10.3, 2.0 Hz, 2H, 2 × CH^{PC}), 3.16 (ddd, J = 12.1, 10.2, 4.6 Hz, 2H, 2 × CH^{PC}), 2.95 (ddd, J = 12.1, 11.4, 2.0 Hz, 2H, 2 × CH^{PC}), 2.85 (ddd, J = 13.0, 10.6, 4.6 Hz, 2H, 2 × CH^{PC}). – **¹³C NMR** (101 MHz, CDCl₃) δ/ppm = 141.30 (C$_q$, 2 × C^{Ar}), 138.66 (C$_q$, 2 × C^{Ar}), 137.46 (+, CH, 2 × C^{Ar}), 134.25 (+, CH, 2 × C^{Ar}), 128.40 (+, CH, 2 × C^{Ar}), 126.87 (C$_q$, 2 × C^{Ar}–Br), 35.51 (–, 2 × CH₂), 32.97 51 (–, 2 × CH₂). – **IR** (ATR): $\tilde{v}$/cm⁻¹ = 2932 (vw), 2849 (vw), 1895 (vw), 1583 (vw), 1532 (vw), 1474 (vw), 1449 (vw), 1432 (vw), 1390 (w), 1313 (vw), 1185 (vw), 1104 (vw), 1030 (w), 947 (vw), 899 (w), 839 (w), 855 (w), 830 (w), 706 (w), 669 (w), 647 (w), 522 (vw), 464 (w), 393 (vw). – **MS** (EI, 70 eV), m/z (%): 364/366/368 (3/6/3) [M]⁺, 184/182 (18/18) [M – C₈H₇Br]⁺, 104 (100) [C₈H₈]⁺. – **HRMS** (EI, C₁₆H₁₄⁷⁹Br₂) calc. 363.9457, found 363.9455. The analytical data matched those reported in the literature. [83,228–230]

(*rac*)-4,12-Dibromo[2.2]paracyclophane/(*rac*)-75

In a 10 mL microwave vessel was placed 4,16-dibromo[2.2]paracyclophane (**71**, 500 mg, 1.37 mmol, 1.00 equiv.) and DMF (1.00 mL). The device was programmed to heat the mixture to 180 °C with a holding time set as 6 min. The maximum pressure for the system was set at 17.2 bar and the power was set at 300 W. After cooling to room temperature, the mixture was diluted with DMF (2 mL) and the insoluble solid was filtrated. The precipitate was collected and the reaction was repeated under the same conditions until all the starting material (5.00 g, 13.7 mmol, 1.00 equiv.) reacted. The combined filtrate was poured into water (75 mL) and exacted with EtOAc (3 × 100 mL). The combined organic phase was washed with water and brine, dried over Na₂SO₄ and concentrated under reduced pressure to give (*rac*)-4,16-dibromo[2.2]paracyclophane **75** as a pale brown power, 3.50 g, 9.65 mmol, 70%.[84]

R_f = 0.68 (*c*-Hex/EtOAc = 9:1). – **¹H NMR** (400 MHz, CDCl₃) δ/ppm = 7.22 (d, J = 1.6 Hz, 2H, 2 × CH^{Ar}), 6.56 (d, J = 7.8 Hz, 2H, 2 × CH^{Ar}), 6.52 (dd, J = 7.9, 1.7 Hz, 2H, 2 × CH^{Ar}), 3.47 (ddd, J = 13.3, 9.6, 2.2 Hz, 2H, 2 × CH^{PC}), 3.10 (ddd, J = 13.0, 9.6, 6.8 Hz, 2H, 2 × CH^{PC}), 3.06–2.94 (m, 2H, 2 × CH^{PC}), 2.82 (ddd, J = 13.3, 10.1, 6.9 Hz, 2H, 2 × CH^{PC}). – **¹³C NMR** (101 MHz, CDCl₃) δ/ppm = 141.35 (C$_q$, 2 × C^{Ar}), 138.79 (C$_q$, 2 × C^{Ar}), 135.08 (+, CH, 2 × C^{Ar}), 132.78 (+, CH, 2 × C^{Ar}), 131.72 (+, CH, 2 × C^{Ar}), 126.70

(C_q, $2 \times C^{Ar}$-Br), 35.89 (–, $2 \times CH_2$), 32.54 89 (–, $2 \times CH_2$). – **IR** (ATR): $\tilde{v}/cm^{-1}$ = 2923 (w), 2848 (w), 1583 (w), 1537 (w), 1474 (w), 1449 (w), 1431 (w), 1391 (m), 1272 (w), 1237 (w), 1201 (w), 1185 (w), 1030 (m), 902 (m), 858 (m), 785 (w), 705 (m), 644 (m), 475 (m). – **MS** (70 eV, EI) *m/z* (%): 368/366/364 (22/43/22) [M]$^+$, 288/286 (13/12) [M+H–Br]$^+$, 184/182 (80/100) [M–C_8H_6Br]$^+$, 104 (68) [C_8H_8]$^+$. – **HRMS** (EI, $C_{16}H_{14}^{79}Br_2$) calc. 363.9462, found 363.9461. The analytical data matched those reported in the literature. [83,84,162,230]

Separation of (*rac*)-4,12-dibromo[2.2]paracyclophane **75** *via* semi-preparative chiral HPLC:

Conditions:

Semi-preparative chiralpak® AZ-H column (20 × 250 nm, particle 5 μm), 100% CH_3CN, 25 mL/min, 25 °C, 254 nm UV detecter, 100 mg racemate per run. t_R (R_P) = 6.97 min, t_R (S_P) = 8.00 min. The spectra of the racemate as well as the enationpure (R_p-4,12-dibromo-[2.2]paracyclophane) **75** are shown in Figure 37.

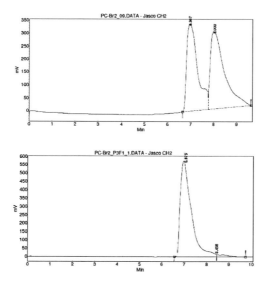

Figure 37. Separation of (*rac*)-4,12-dibromo[2.2]paracyclophane/(*rac*)-**75** *via* semi-preparative HPLC.

(R_p)-4,12-Dicarboxy[2.2]paracyclophane (R_p-146)

To a solution of (R_p)-4,12-dibromo[2.2]paracyclophane (R_p-75, 1.50 g, 4.12 mmol, 1.00 equiv.) in abs. THF (50 mL) was added 9.71 mL of *t*-butyllithium (1.7 M in pentane, 15.4 mmol, 4.00 equiv.) dropwise at –78 °C. After stirring at – 78 °C for 3 h, CO_2 was bubbled through the solution *via* a long needle under stirring for 2 h. The reaction mixture was then quenched with water and extracted with 1 M NaOH solution (2 × 100 mL) until all the gas was consumed. The water phases were combined, washed with CH_2Cl_2 (50 mL) and acidified with 6 M HCl until the solution showed acid. The precipitate was filtrated, washed with water and CH_2Cl_2. The product R_p-146 was obtained after drying under high vacuum as white powder, 640 mg, 3.14 mmol, 52%.

$[\alpha]_D^{20}$ = –134 (c = 0.00203, EtOH). – ^{1}H NMR (400 MHz, DMSO-d_6) δ/ppm = 12.45 (s, 2H, 2 × COOH), 7.04 (d, J = 2.0 Hz, 2H, 2 × CH^{Ar}), 6.78 (dd, J = 7.8, 1.9 Hz, 2H, 2 × CH^{Ar}), 6.60 (d, J = 7.8 Hz, 2H, 2 × CH^{Ar}), 4.04–3.88 (m, 2H, 2 × CH^{Ar}), 3.15 (dd, J = 12.5, 9.8 Hz, 2H, 2 × CH^{Ar}), 2.98 (ddd, J = 12.5, 9.6, 7.3 Hz, 2H, CH^{Ar}), 2.81 (ddd, J = 12.3, 9.8, 7.3 Hz, 2H, 2 × CH^{Ar}). – ^{13}C NMR (101 MHz, DMSO-d_6) δ/ppm = 167.76 (C_q, 2 × COOH), 141.98 (C_q, 2 × C^{Ar}), 139.76 (C_q, 2 × C^{Ar}), 136.16 (+, CH, 2 × C^{Ar}), 135.98 (+, CH, 2 × C^{Ar}), 133.30 (+, CH, 2 × C^{Ar}), 130.76 (C_q, 2 × C^{Ar}-COOH), 35.30 (–, CH_2, 2 × C^{PC}), 33.78 (–, CH_2, 2 × C^{PC}). – IR (ATR): $\tilde{v}$/cm^{-1} = 2925 (w), 1674 (w), 1592 (w), 1556 (w), 1489 (vw), 1422 (w), 1300 (w), 1273 (w), 1203 (w), 1074 (w), 909 (w), 850 (vw), 797 (vw), 759 (vw), 717 (vw), 664 (w), 631 (w), 555 (vw), 518 (w), 426 (vw). – MS (70 eV, EI) *m/z (%)*: 296 (27) [M]$^+$, 278 (100) [M–H_2O]$^+$, 148 (83) [M–$C_9H_8O_2$]$^+$. – HRMS (EI, $C_{18}H_{16}O_4$) calc. 296.1049, found 296.1049. The analytical data match those reported in literature.[158]

(R_p, S)-4,12-Bis(4'-isopropyloxazolin-2'yl)[2.2]paracyclophane / (R_p,S)-144a

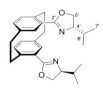

Thionyl chloride (1.0 mL) was added to (R_p,S)-4,12-dicarboxyl[2.2]paracyclophane (R_p-146, 250 mg, 0.840 mmol, 1.00 equiv.) and the resulting mixture was stirred at 100 °C for 90 min. After cooling to room temperature the excess thionyl chloride was removed under vacuum, the final traces were washed with toluene (2 × 2 mL) and removed under vacuum. The resulting crude acetyl chloride was dissolved in CH_2Cl_2 (5 mL) and cooled to 0 °C. A solution of S-Valinol (145a, 0.350 g, 3.36 mmol, 4.00 equiv.) and Et$_3$N (0.54 mL, 0.42 g, 4.20 mmol, 5.00 equiv.) in CH_2Cl_2 (1.0 mL) was added, the reaction mixture was allowed to warm to room temperature and stirred for 24 h. 10 mL of CH_2Cl_2 was then added and the solution was washed with aq. NaHCO$_3$ solution (3.5% w/v, 2 × 10 mL) and brine (20 mL). The organic phase was dried over MgSO$_4$, filtered, concentrated and dried under vacuum to give the crude amide as light brown solid.

The crude amide was dissolved in CH$_3$CN (5.0 mL), PPh$_3$ (0.66 g, 2.52 mmol, 3.00 equiv.), CCl$_4$ (0.770 mL, 1.23 g, 7.98 mmol, 9.50 equiv.) and Et$_3$N (0.970 mL, 0.760 g, 7.56 mmol, 9.00 equiv.) were added subsequently. After stirring at room temperature overnight, the solvent was removed under reduced pressure, the resulting mixture was dissolved in CH$_2$Cl$_2$ and washed with H$_2$O (2 × 10 mL), the combined organic phase was washed with brine, dried over Na$_2$SO$_4$, filtrated and concentrated under vacuum. The crude was purified *via* column chromatography (*c*-Hex/EtOAc = 9:1) to give the title compound (*R*$_p$,*S*)-**144a** as a pale yellow solid, 0.150 g, 0.350 mmol, 42%.

R$_f$ = 0.34   (*c*-Hex/EtOAc = 9:1). – **^{1}H   NMR**   (400 MHz,   CDCl$_3$)   δ/ppm = 7.09 (d, *J* = 1.9 Hz, 2H, 2 × C*H*^{Ar}), 6.62 (dd, *J* = 7.9, 1.9 Hz, 2H, 2 × C*H*^{Ar}), 6.54 (d, *J* = 7.8 Hz, 2H, 2 × C*H*^{Ar}), 4.37 (ddd, *J* = 11.2, 9.5, 2.0 Hz, 2H, 2 × C*H*^{PC}), 4.30 (dd, *J* = 5.8, 2.2 Hz, 2H, 2 × C*H*$^{5'}$), 4.04 (dd, *J* = 8.6, 6.7 Hz, 2H, 2 × C*H*$^{4'}$), 4.05–3.92 (m, 2H, 2 × C*H*$^{5'}$), 3.24–3.16 (m, 2H, 2 × C*H*^{PC}), 3.16–3.07 (m, 2H, 2 × C*H*^{PC}), 2.82 (ddd, *J* = 12.6, 10.0, 7.1 Hz, 2H, 2 × C*H*^{PC}), 1.95 (hept, *J* = 6.7 Hz, 2H, 2 × C*H*$^{6'}$), 1.20 (d, *J* = 6.7 Hz, 6H, C*H*$^{7'}$), 1.06 (d, *J* = 6.7 Hz, 6H, C*H*$^{7'}$). – **^{13}C NMR** (101 MHz, CDCl$_3$) δ/ppm = 162.91 (C$_q$, 2 × C$^{2'}$), 141.08 (C$_q$, 2 × C^{Ar}), 140.12 (C$_q$, 2 × C^{Ar}), 135.81 (+, CH, 2 × C^{Ar}), 134.87 (+, CH, 2 × C^{Ar}), 132.33 (+, CH, 2 × C^{Ar}), 128.24 (C$_q$, 2 × C^{Ar}), 73.84 (+, CH, 2 × C$^{4'}$), 69.39 (–, CH$_2$, 2 × C$^{5'}$), 35.84 (–, CH$_2$, 2 × C^{PC}), 33.62 (+, CH, 2 × C$^{6'}$), 33.61 (–, CH$_2$, 2 × C^{PC}), 19.76 (+, CH$_3$, 2 × C$^{7'}$), 19.23 (+, CH$_3$, 2 × C$^{7'}$). – **IR** (ATR): $\tilde{v}$/cm^{-1} = 2955 (w), 1637 (m), 1590 (w), 1492 (w), 1468 (w), 1429 (w), 1384 (m), 1346 (w), 1303 (w), 1275 (w), 1258 (w), 1191 (w), 1172 (w), 1137 (w), 1115 (w), 1053 (m), 1026 (w), 984 (m), 933 (w), 907 (m), 889 (w), 822 (w), 749 (w), 694 (w), 674 (w), 643 (w), 514 (w), 482 (vw), 389 (vw). – **MS** (FAB, 3-NBA), m/z (%): 431 (100) [M + H]$^+$, 500/488 (9/9) [C$_{14}$H$_{17}$NO$_2$ + H]$^+$. – **HRMS** (FAB, C$_{28}$H$_{35}$O$_2$N$_2$, [M+H]$^+$): calc. 431.2699, found 431.2701. The analytical data match those reported in literature.[231]

(R_p,S)-N^4,N^{12}-Bis((S)-1′-hydroxy-3′,3′-dimethylbutan-2-yl)[2.2]paracyclophane-4,12-dicarboxy amide / (R_p,S)-148b

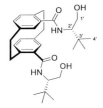 Thionyl chloride (2.0 mL) was added to (R_p)-4,12-dicarboxyl[2.2]paracyclophane (R_p-**146**, 0.150 g, 0.510 mmol, 1.00 equiv.), after stirring at room temperature for 10 min, the mixture was heated to 100 °C and stirred at this temperature for 90 min. The excess thionyl chloride was removed by evaporation, the final traces were washed with toluene (2 × 2 mL). After drying under vacuum, the resulting crude acetyl chloride was dissolved in abs. CH$_2$Cl$_2$ (5 mL) and cooled to 0 °C. A solution of (S)-(+)-*tert*-leucinol (**145b**, 0.229 g, 2.04 mmol, 4.00 equiv.) and abs. Et$_3$N (0.260 g, 0.360 mL, 2.55 mmol, 5.00 equiv.) in CH$_2$Cl$_2$ (1.0 mL) was added and the reaction mixture allowed to warm to room temperature and stirred for 24 h. Water was then added (10 mL) and extracted with CH$_2$Cl$_2$ (3 × 10 mL), the combined organic phase was washed with sat. NaHCO$_3$ solution and brine (20 mL). The organic phase was dried over MgSO$_4$, filtrated, concentrated and dried under vacuum. The crude was purified *via* column chromatography (CH$_2$Cl$_2$/MeOH = 98:2 → 95:5) to give the title compound (R_p,S)-**148b** as colorless solid, 152 mg, 0.310 mmol, 60%.

R_f = 0.24 (CH$_2$Cl$_2$/MeOH = 98:2). – **^{1}H NMR** (400 MHz, CDCl$_3$) δ/ppm = 7.21 (d, J = 9.3 Hz, 2H, 2 × NH), 6.99 (d, J = 1.7 Hz, 2H, 2 × CH^{Ar}), 6.63 (dd, J = 7.8, 1.7 Hz, 2H, 2 × CH^{Ar}), 6.58 (d, J = 7.8 Hz, 2H, 2 × CH^{Ar}), 4.02 (ddd, J = 9.3, 8.2, 3.3 Hz, 2H, 2 × CH^{PC}), 3.91 (dd, J = 11.4, 3.3 Hz, 2H, 2 × C$H^{1'}$), 3.83–3.72 (m, 1H, C$H^{2'}$), 3.68 (dd, J = 11.4, 8.2 Hz, 2H, 2 × C$H^{1'}$), 3.19–3.02 (m, 1H, C$H^{2'}$), 2.99–2.91 (m, 2H, 2 × CH^{PC}), 2.91–2.76 (m, 4H, 4 × CH^{PC}), 1.04 (s, 18H, C$H^{4'}$). – **^{13}C NMR** (101 MHz, CDCl$_3$) δ/ppm = 170.97 (C$_q$, 2 × C=O), 140.27 (C$_q$, 2 × C^{Ar}), 138.49 (C$_q$, 2 × C^{Ar}), 136.17 (C$_q$, 2 × C^{Ar}), 135.92 (+, CH, 2 × C^{Ar}), 134.71 (+, CH, 2 × C^{Ar}), 130.30 (+, CH, 2 × C^{Ar}), 63.06 (–, CH$_2$, 2 × $C^{1'}$), 60.09 (+, CH, 2 × $C^{2'}$), 34.71 (–, CH$_2$, 2 × C^{PC}), 34.65 (–, CH$_2$, 2 × C^{PC}), 34.42 (C$_q$, 2 × $C^{3'}$), 27.37 (+, CH$_3$, 6 × $C^{4'}$). – **IR** (ATR): $\tilde{v}$/cm^{-1} = 3232 (vw), 2961 (w), 1624 (w), 1589 (w), 1536 (w), 1474 (w), 1436 (w), 1397 (w), 1366 (w), 1351 (w), 1306 (w), 1237 (vw), 1205 (vw), 1110 (vw), 1091 (w), 1047 (w), 1020 (w), 998 (w), 967 (vw), 908 (vw), 894 (vw), 825 (vw), 807 (vw), 786 (w), 730 (w), 628 (w), 503 (w), 474 (vw).– **MS** (FAB, 3-NBA), m/z (%): 495 (100) [M + H]$^+$, 248 (8) [C$_{15}$H$_{21}$NO$_2$ + H]$^+$. – **HRMS** (FAB, C$_{30}$H$_{43}$O$_4$N$_2$), [M+H]$^+$): calc 495.3223, found 495.3225.

(R_p, S)-4,12-Bis(4'-*tert*butyloxazolin-2'yl)[2.2]paracyclophane / (R_p,S)-144b

To a solution of (R_p,S)-N^4,N^{12}-bis$((S)$-1'-hydroxy-3',3'-dimethylbutan-2-yl)[2.2]paracyclophane-4,12-dicarboxy amide $((R_p,S)$-**148b**, 152 mg, 0.307 mmol, 1.00 equiv.) and PPh₃ (282 mg, 1.08 mmol, 3.50 equiv.) in abs. CH₃CN (8.00 mL) was added Et₃N (0.385 mL, 280 mg, 2.76 mmol, 9.00 equiv.) and CCl₄ (0.281 mL, 449 mg, 2.92 mmol, 9.50 equiv.) under argon atmosphere. After stirring at room temperature over night, the solvent was removed under vacuum, the resulting crude was dissolved in CH₂Cl₂ and washed with brine, the organic phase was dried over Na₂SO₄, filtrated and concentrated under vacuum. The resulting mixture was purified *via* column chromatography (*c*-Hex/EtOAc = 9:1) to give title product (R_p,S)-**144b** as colorless solid, 104 mg, 0.227 mmol, 74%.

R_f = 0.36 (*c*-Hex/EtOAc = 9:1). – **¹H NMR** (400 MHz, CDCl₃) δ/ppm = 7.12 (d, *J* = 1.9 Hz, 2H, 2 × CH^{Ar}), 6.64 (dd, *J* = 7.8, 1.9 Hz, 2H, 2 × CH^{Ar}), 6.55 (d, *J* = 7.8 Hz, 2H, 2 × CH^{Ar}), 4.33–4.24 (m, 2H, 2 × $CH^{5'}$), 4.20 (td, *J* = 8.8, 3.7 Hz, 2H, 2 × $CH^{4'}$), 4.17–4.08 (m, 4H, 2 × $CH^{5'}$ + 2 × CH^{PC}), 3.21–3.03 (m, 4H, 4 × CH^{PC}), 2.86–2.68 (m, 2H, 2 × CH^{PC}), 0.99 (s, 18H, $CH^{7'}$). – **¹³C NMR** (101 MHz, CDCl₃) δ/ppm = 162.91 (C_q, 2 × $C^{2'}$), 141.03 (C_q, 2 × C^{Ar}), 140.27 (C_q, 2 × C^{Ar}), 135.64 (+, CH, 2 × C^{Ar}), 134.71 (+, CH, 2 × C^{Ar}), 132.25 (+, CH, 2 × C^{Ar}), 128.02 (C_q, 2 × C^{Ar}), 74.20 (+, CH, 2 × $C^{4'}$), 67.78 (–, CH₂, 2 × $C^{5'}$), 36.25 (–, CH₂, 2 × C^{PC}), 34.11 (–, CH₂, 2 × C^{PC}), 34.06 (C_q, 2 × $C^{6'}$), 26.16 (+, CH₃, 6 × $C^{7'}$). – **IR** (ATR): $\tilde{v}$/cm⁻¹ = 2951 (w), 2866 (w), 1638 (m), 1590 (w), 1477 (w), 1392 (w), 1350 (w), 1333 (w), 1303 (w), 1257 (w), 1191 (w), 1172 (w), 1113 (w), 1067 (w), 1047 (w), 1024 (w), 979 (m), 930 (w), 906 (w), 819 (w), 791 (w), 719 (w), 679 (w), 632 (w), 544 (vw), 513 (w). – **MS** (FAB, 3-NBA), m/z (%): 459 (82) [M + H]⁺, 230 (75) [C₁₅H₁₉NO + H]⁺. – **HRMS** (FAB, C₃₀H₃₉O₂N₂, [M+H]⁺): calc. 459.3012, found 459.3011.

$(R_p,$ $S)$-N^4, N^{12}-Bis $((S)$-2'-hxdroxy-1'-phenylehtyl)[2.2]paracyclophane-4,12-dicarboxy amide / (R_p,S)-148c

Thionyl chloride (2.0 mL) was added to (R_p)-4,12-dicarboxyl[2.2]paracyclophane (R_p-**146**, 0.150 g, 0.510 mmol, 1.00 equiv.), after stirring at room temperature for 10 min, the mixture was heated to 100 °C and stirred under this temperature for 90 min. The excess thionyl chloride was removed by evaporation and the final traces were washed with toluene (2 × 2 mL). After drying under vacuum, the resulting crude acid chloride was dissolved in abs. CH₂Cl₂ (5 mL) and cooled to 0 °C. A solution of

(S)-(+)-phenylglycinol (**145c**, 0.280 g, 2.04 mmol, 4.00 equiv.) and abs. Et₃N (0.360 mL, 0.260 g, 2.55 mmol, 5.00 equiv.) in CH₂Cl₂ (1 mL) was added and the reaction mixture allowed to warm to room temperature and stirred for 24 h. Water (10 mL) was then added, the water phase was extracted with CH₂Cl₂ (3 × 10 mL) and the combined organic phase was washed with sat. NaHCO₃ solution and brine (20 mL). The organic phase was dried over MgSO₄, filtered, concentrated and dried in vacuum. The crude was purified *via* column chromatography (CH₂Cl₂/MeOH = 98:2 → 95:5) to give the title compound (R_p,S)-**148c** as colorless solid, 232 mg, 0.440 mmol, 85%.

R_f = 0.28 (CH₂Cl₂/MeOH = 98:2). – **¹H NMR** (400 MHz, CDCl₃) δ = 8.11 (s, 2H, 2 × NH), 7.52–7.29 (m, 10H, C$H^{2''+3''+4''}$), 6.86 (s, 2H, 2 × CH^{Ar}), 6.60 (d, J = 7.5 Hz, 2H, 2 × CH^{Ar}), 6.53 (d, J = 7.4 Hz, 2H, 2 × CH^{Ar}), 5.22 (d, J = 6.1 Hz, 2H, 2 × C$H^{2'}$), 4.00–3.88 (m, 4H, 4 × CH^{PC}), 3.69 (t, J = 11.2 Hz, 2H, 2 × C$H^{1'}$), 3.06 (t, J = 11.0 Hz, 2H, 2 × C$H^{1'}$), 2.88–2.62 (m, 4H, 4 × CH^{PC}). – **¹³C NMR** (101 MHz, CDCl₃) δ/ppm = 170.12 (C_q, 2 × C=O), 140.13 (C_q, 2 × $C^{1''}$), 139.32 (C_q, 2 × C^{Ar}), 138.59 (C_q, 2 × C^{Ar}), 136.16 (C_q, 2 × C^{Ar}), 135.84 (+, CH, 2 × C^{Ar}), 134.97 (+, CH, 2 × C^{Ar}), 130.44 (+, CH, 2 × C^{Ar}), 128.88 (+, CH, 4 × $C^{3''}$), 127.90 (+, CH, 2 × $C^{4''}$), 127.21 (+, CH, 4 × $C^{2''}$), 66.31 (–, CH₂, 2 × $C^{2'}$), 56.88 (+, CH, 2 × $C^{1'}$), 34.85 (–, CH₂, 2 × C^{PC}), 34.68 (–, CH₂, 2 × C^{PC}). – **IR** (ATR): ṽ/cm⁻¹ = 3212 (vw), 2925 (vw), 1634 (w), 1613 (w), 1589 (w), 1525 (w), 1494 (w), 1455 (w), 1364 (vw), 1301 (vw), 1274 (vw), 1206 (vw), 1067 (w), 1030 (vw), 902 (vw), 833 (vw), 763 (vw), 739 (w), 695 (w), 631 (w), 561 (vw), 546 (vw), 531 (w), 509 (w), 483 (w). – **MS** (FAB, 3-NBA), m/z (%): 535 (100) [M + H]⁺, 278 (26).– **HRMS** (FAB, C₃₄H₃₅O₄N₂, [M+H]⁺): calc. 535.2597, found 535.2598.

(R_p,S)-4,12-Bis(1'-phenyloxazolin-2'yl)[2.2]paracyclophane/ (R_p,S)-144c

To a solution of (R_p,S)-N^4,N^{12}-bis((S)-2'-hydroxy-1'-phenylethyl)[2.2]paracyclophane-4,12-dicarboxyamide ((R_p,S)-**148c**, 200 mg, 0.374 mmol, 1.00 equiv.) and PPh₃ (344 mg, 1.31 mmol, 3.50 equiv.) in abs. 10 mL of CH₃CN was added Et₃N (0.469 mL, 341 mg, 3.37 mmol, 9.00 equiv.) and CCl₄ (0.343 mL, 547 mg, 3.55 mmol, 9.50 equiv.) under argon atmosphere. After stirring at room temperature overnight, the solvent was removed under vacuum, the resulting crude was dissolved in CH₂Cl₂ and washed with brine. The organic phase was dried over Na₂SO₄, filtrated and concentrated under vacuum, the resulting mixture was purified *via* column chromatography (*c*-Hex/EtOAc = 9:1) to give title product (R_p,S)-**144c** as colorless solid, 177 mg, 0.355 mmol, 95%.

R_f = 0.14 (c-Hex/EtOAc = 9:1). – ^{1}H NMR (500 MHz, CDCl$_3$) δ/ppm = 7.40–7.29 (m, 12H, C$H^{7+8'+9}$ + 2 × CH^{Ar}), 6.70 (dd, J = 7.9, 1.9 Hz, 2H, 2 × CH^{Ar}), 6.61 (d, J = 7.9 Hz, 2H, 2 × CH^{Ar}), 5.47 (dd, J = 10.1, 8.2 Hz, 2H, 2 × C$H^{5'}$), 4.65 (dd, J = 10.1, 8.2 Hz, 2H, 2 × C$H^{5'}$), 4.44–4.25 (m, 2H, 2 × CH^{PC}), 4.13 (t, J = 8.2 Hz, 2H, 2 × C$H^{4'}$), 3.22–3.14 (m, 4H, 4 × CH^{PC}), 2.89–2.79 (m, 2H, 2 × CH^{PC}). – ^{13}C NMR (126 MHz, CDCl$_3$) δ/ppm = 164.62 (C$_q$, 2 × C$^{2'}$), 143.04 (C$_q$, 2 × C$^{5'}$), 141.49 (C$_q$, 2 × C^{Ar}), 140.34 (C$_q$, 2 × C^{Ar}), 135.97 (+, CH, 2 × C^{Ar}), 135.16 (+, CH, 2 × C^{Ar}), 132.85 (+, CH, 2 × C^{Ar}), 128.80 (+, CH, 4 × C$^{8'}$), 128.59 (C$_q$, CH, 2 × C^{Ar}), 127.59 (+, CH, 2 × C$^{9'}$), 126.95 (+, CH, 4 × C$^{7'}$), 73.94 (+, CH, 2 × C$^{4'}$), 70.75 (–, CH$_2$, 2 × C$^{5'}$), 36.44 (–, CH$_2$, 2 × C^{PC}), 34.26 (–, CH$_2$, 2 × C^{PC}). – IR (ATR): $\tilde{v}$/cm^{-1} = 2922 (w), 1630 (m). 1589 (w), 1493 (w), 1448 (w), 1349 (w), 1296 (w), 1274 (w), 1245 (w), 1191 (w), 1172 (w), 1136 (vw), 1116 (vw), 1050 (w), 986 (w), 961 (w), 927 (w), 902 (w), 887 (w), 823 (w), 750 (w), 697 (w), 639 (w), 523 (w), 388 (vw). – MS (FAB, 3-NBA), m/z (%): 499 (100) [M + H]$^+$, 250 (34) [C$_{17}$H$_{15}$NO + H]$^+$. – HRMS (FAB, C$_{34}$H$_{31}$O$_2$N$_2$, [M+H]$^+$): calc 499.2386, found 499.2386.

(rac)-4-Nitro[2.2]paracyclophane (159)

A stirred solution of [2.2]paracyclophane (59, 416 mg, 2.00 mmol, 1.00 equiv.) in 10 mL of glacial AcOH was heated to 90 °C for 30 min. The suspension was then cooled to 60 °C rapidly fuming HNO$_3$ (1.26 g, 0.85 mL, 20.0 mmol, 2.00 equiv.) was added carefully. The mixture was stirred until a solution was obtained (30 – 60 s). The solution was immediately poured onto ice and the aqueous phase was extracted with CH$_2$Cl$_2$ (3 × 20 mL). The combined organic phase was washed with aq. 1 M NaOH (30 mL), dried over Na$_2$SO$_4$, filtered, and the solvent was removed under reduced pressure. The residue was subjected to flash chromatography on silica gel (c-Hex/EtOAc = 4:1) to obtain (rac)-4-nitro-[2.2]para-cyclophane 159 as light yellow solid, 298 mg, 1.18 mmol, 59%.

R_f = 0.44 (c-Hex/EtOAc = 5:1). – ^{1}H NMR (400 MHz, CDCl$_3$) δ/ppm = 7.22 (d, J = 1.9 Hz, 1H, CH^{Ar}), 6.79 (dd, J = 7.9, 1.9 Hz, 1H, CH^{Ar}), 6.64–6.61 (m, 2H, 2 × CH^{Ar}), 6.57 (dd, J = 4.6, 1.8 Hz, 2H, 2 × CH^{Ar}), 6.48 (dd, J = 7.9, 1.7 Hz, 1H, CH^{Ar}), 4.03 (ddd, J = 13.2, 9.3, 2.1 Hz, 1H, CH^{PC}), 3.26–3.21 (m, 1H, CH^{PC}), 3.17 (ddt, J = 13.0, 5.8, 2.6 Hz, 4H, 4 × CH^{PC}), 3.12–3.01 (m, 2H, CH^{PC}), 2.90 (ddd, J = 13.2, 10.0, 7.2 Hz, 1H, CH^{PC}). – ^{13}C NMR (101 MHz, CDCl$_3$) δ/ppm = 149.40 (C$_q$, C^{Ar}-NO$_2$), 142.18 (C$_q$, C^{Ar}), 139.89 (C$_q$, C^{Ar}), 139.42 (C$_q$, C^{Ar}), 137.88 (+, CH, C^{Ar}), 137.45 (+, CH, C^{Ar}), 136.58 (C$_q$, C^{Ar}), 133.30 (+, CH, C^{Ar}), 133.23 (+, CH, C^{Ar}), 132.52 (+, CH, C^{Ar}), 130.09 (+, CH, C^{Ar}), 129.67 (+, CH, C^{Ar}), 36.14 (–, CH$_2$, C^{PC}), 35.10 (–, CH$_2$, C^{PC}), 34.92 (–, CH$_2$, C^{PC}), 34.57 (–, CH$_2$, C^{PC}). – IR (ATR): $\tilde{v}$/cm^{-1} = 2922 (w), 2850 (vw), 1593 (vw), 1550 (m), 1515 (w), 1482 (w), 1451 (w), 14365 (w), 1410 (vw), 1331 (m), 1201 (w), 1180 (w),

1094 (w), 930 (vw), 904 (w), 870 (w), 805 (w), 788 (w), 757 (w), 703 (w), 673 (w), 634 (m), 582 (vw), 507 (w), 464 (vw). – **MS** (EI, 70 eV), m/z (%): 253 (68) $[M]^+$, 207 (14) $[M - NO_2]^+$, 104 (100) $[C_8H_8]^+$. – **HRMS** (EI, $C_{16}H_{15}O_2N_1$) calc. 253.1097, found 253.1097. The analytical data match those reported in the literature.[77,194,232]

(*rac*)-13-Nitro-4-bromo[2.2]paracyclophane (160)

 A solution of Br_2 (0.100 mL, 2.00 mmol, 1.00 equiv.) in CCl_4 (8 mL) was prepared. A suspension of iron powder (20 mg) in 1 mL of the Br_2/CCl_4 solution was stirred at room temperature for 1 h. CH_2Cl_2 (20 mL) and 4-nitro[2.2]paracyclophane (**159**, 506 mg, 2.00 mmol, 1.00 equiv.) were added subsequently. The remaining Br_2/CCl_4 solution was added dropwise and the solution was stirred at room temperature overnight followed by TLC check. After the starting material was consumed, the reaction mixture was washed with aqueous sat. $NaHCO_3$ (2 × 10 mL), the combined organic phase was washed with brine, dried over Na_2SO_4 and concentrated under vacuum. The crude was purified *via* column chromatography (*c*-Hex/EtOAc = 4:1) to give **160** as yellow solid, 145 mg, 0.440 mmol, 22%.

R_f = 0.33 (*c*-Hex/EtOAc =4:1). – **¹H NMR** (500 MHz, CDCl₃) δ/ppm = 7.57 (d, *J* = 1.7 Hz, 1H, C*H*^{Ar}), 6.77 (dd, *J* = 7.8, 1.8 Hz, 1H, C*H*^{Ar}), 6.70 (d, *J* = 1.6 Hz, 1H, C*H*^{Ar}), 6.66–6.57 (m, 3H, 3 × C*H*^{Ar}), 4.38 (ddd, *J* = 13.8, 9.7, 4.2 Hz, 1H, C*H*^{PC}), 3.66 (ddd, *J* = 13.7, 9.8, 4.0 Hz, 1H, C*H*^{PC}), 3.30–2.71 (m, 6H, 6 × C*H*^{PC}). – **¹³C NMR** (126 MHz, CDCl₃) δ/ppm = 147.54 (C_q, C^{Ar}-NO₂), 140.98 (C_q, C^{Ar}), 140.96 (C_q, C^{Ar}), 138.61 (C_q, C^{Ar}), 137.93 (+, CH, C^{Ar}), 137.56 (+, CH, C^{Ar}), 136.31 (C_q, C^{Ar}), 136.02 (+, CH, C^{Ar}), 134.66 (+, CH, C^{Ar}), 131.32 (+, CH, C^{Ar}), 127.89 (+, CH, C^{Ar}), 127.29 (C_q, C^{Ar}-Br), 35.30 (–, CH₂, C^{PC}), 34.52 (–, CH₂, C^{PC}), 34.26 (–, CH₂, C^{PC}), 32.58 (–, CH₂, C^{PC}). – **IR** (ATR): $\tilde{v}$/cm⁻¹ = 2931 (w), 1584 (vw), 1515 (m), 1474 (w), 1449 (w), 1391 (w), 1333 (m), 1167 (w), 1031 (w), 1390 (w), 959 (w), 929 (vw), 909 (w), 882 (w), 843 (w), 798 (w), 750 (w), 708 (w), 690 (w), 666 (w), 645 (w), 521 (w), 477 (vw).– **MS** (EI, 70 eV), m/z (%): 333/331 (21/16) $[M]^+$, 184/182 (100/98) $[M - NO_2 - C_8H_8]^+$, 104 (12) $[C_8H_8]^+$. – **HRMS** (EI, $C_{16}H_{14}O_2{}^{79}Br_1$) calc. 331.0202, found 331.0201. The analytical data match those reported in the literature.[230,232]

(*rac*)-13-Amino-4-bromo[2.2]paracyclophane (161)

To a solution of EtOH (4 mL) and H₂O (4 mL) was suspended (*rac*)-4-bromo-13-nitro[2.2]paracyclophane (**160**, 66.2 mg, 0.200 mmol, 1.0 equiv.) and the mixture was stirred at room temperature for 1 h. Iron powder (134 mg, 2.4 mmol, 12.0 equiv.) was added and the mixture was heated to reflux. 0.4 mL of concentrated HCl was added carefully and the mixture was stirred under reflux until no starting material left. The reaction mixture was allowed to cool to room temperatur and poured onto ice/sat. aq NaHCO₃ (500 mL). The mixture was extracted with EtOAc (3 ×20 mL). The combined organic phase was washed with brine, dried over Na₂SO₄, filtered, and the solvent was removed under reduced pressure. The residue was subjected to flash chromatography on silica gel (*c*-Hex/EtOAc = 4:1) to give (*rac*)-13-amino-4-bromo[2.2]paracyclophane **161** as colorless solid, 56.0 mg, 0.190 mmol, 93%.

R_f = 0:49 (*c*-Hex/EtOAc = 4:1). – **¹H NMR** (500 MHz, CDCl₃) δ/ppm = 6.83 (d, J = 1.7 Hz, 1H, CH^{Ar}), 6.49 (dd, J = 7.8, 1.7 Hz, 1H, CH^{Ar}), 6.44 (d, J = 7.7 Hz, 1H, CH^{Ar}), 6.39 (d, J = 7.6 Hz, 1H, CH^{Ar}), 6.22–5.98 (m, 1H, CH^{Ar}), 5.75 (s, 1H, CH^{Ar}), 4.01 (brs, 2H, NH_2), 3.70 (ddd, J = 13.8, 9.7, 5.7 Hz, 1H, CH^{PC}), 3.36 (ddd, J = 14.2, 9.8, 2.5 Hz, 1H, CH^{PC}), 3.11 (ddd, J = 13.6, 10.6, 2.5 Hz, 1H, CH^{PC}), 3.06–2.99 (m, 1H, CH^{PC}), 2.97–2.81 (m, 4H, 4 × CH^{PC}).– **¹³C NMR** (126 MHz, CDCl₃) δ/ppm = 146.27 (C$_q$, C^{Ar}-NH₂), 141.10 (C$_q$, C^{Ar}), 140.75 (C$_q$, C^{Ar}), 138.27 (C$_q$, C^{Ar}), 135.55 (+, CH, C^{Ar}), 135.40 (+, CH, C^{Ar}), 135.32 (+, CH, C^{Ar}), 132.82 (+, CH, C^{Ar}), 124.06 (C$_q$, C^{Ar}), 123.64 (+, CH, C^{Ar}), 122.67 (C$_q$, C^{Ar}-Br), 121.45 (+, CH, C^{Ar}), 34.96 (–, CH₂, C^{PC}), 34.81 (–, CH₂, C^{PC}), 33.32 (–, CH₂, C^{PC}), 31.50 (–, CH₂, C^{PC}).– **IR** (ATR): $\tilde{v}$/cm⁻¹ = 3394 (w), 2922 (w), 2851 (w), 1609 (w), 1586 (w), 1561 (w), 1495 (w), 1477 (w), 1422 (w), 1390 (w), 1286 (w), 1164 (w), 1031 (w), 930 (vw), 902 (w), 879 (w), 862 (w), 822 (w), 763 (w), 719 (w), 704 (w), 677 (w), 657 (w), 605 (vw), 513 (w), 456 (w), 434 (w), 387 (w). – **MS** (FAB, 3-NBA), m/z (%): 303/301 (100/100) [M]⁺, 119 (61) [C₈H₉N]⁺. – **HRMS** (FAB, C₁₆H₁₆N₁⁷⁹Br₁, [M]⁺): calc. 301.0456, found 301.0467. The analytical data match those reported in the literature.[230,232,233]

(*rac*)-13, 13-Dimethylamino-4-bromo[2.2]paracyclophane (162)

A round-bottomed flask was charged with (*rac*)-13-amino-4-bromo [2.2]paracyclophane (**161**, 200 mg, 0.680 mmol, 1.00 equiv.), iodomethane (100 μL, 290 mg, 2.55 mmol, 3.00 equiv.), sodium carbonate (180 mg, 1.70 mmol, 2.00 equiv.) and DMF (8.00 mL) and then purged with argon. The mixture was heated to 50 °C and stirred overnight until the starting material was completely consumed. The reaction mixture was diluted with 5 mL of water and extracted with Et$_2$O (3 × 10 mL), the combined organic layer was dried over anhydrous Na$_2$SO$_4$, filtered and concentrated in vacuo. The crude product was purified *via* column chromatography (*c*-Hex/EtOAc = 5:1) to give (*rac*)-13-dimethylamino-4-bromo[2.2]para-cyclophane **162** as light yellow solid, 196 mg, 0.590 mmol, 88%.

R$_f$ = 0.64 (*c*-Hex/EtOAc = 4:1) – **¹H NMR** (400 MHz, CDCl$_3$) δ/ppm = 6.62 (s, 1H, C*H*^{Ar}), 6.51 (s, 2H, 2 × C*H*^{Ar}), 6.48 (d, *J* = 7.5 Hz, 1H, C*H*^{Ar}), 6.22–6.21 (m, 1H, C*H*^{Ar}), 5.80 (s, 1H, C*H*^{Ar}), 3.64 (d, *J* = 10.9 Hz, 1H, *H*^{PC}), 3.55 (d, *J* = 12.0 Hz, 1H, C*H*^{PC}), 3.09–2.86 (m, 6H, 6 × C*H*^{PC}), 2.74 (s, 6H, N(C*H*$_3$)$_2$). – **¹³C NMR** (101 MHz, CDCl$_3$) δ/ppm = 152.47 (C$_q$, *C*^{Ar}-NH$_2$), 140.72 (C$_q$, *C*^{Ar}), 140.14 (C$_q$, *C*^{Ar}), 139.01 (C$_q$, *C*^{Ar}), 136.27 (+, CH, *C*^{Ar}), 135.33 (+, CH, *C*^{Ar}), 135.06 (+, CH, *C*^{Ar}), 131.83 (+, CH, *C*^{Ar}), 129.28 (C$_q$, *C*^{Ar}), 125.86 (+, CH, *C*^{Ar}), 123.36 (C$_q$, *C*^{Ar}-Br), 119.01 (+, CH, *C*^{Ar}), 44.14 (+, CH$_3$, 2 × NCH$_3$), 35.24 (–, CH$_2$, *C*^{PC}), 34.94 (–, CH$_2$, *C*^{PC}), 34.71 (–, CH$_2$, *C*^{PC}), 33.73 (–, CH$_2$, *C*^{PC}). – **IR** (ATR): ṽ/cm^{-1} = 2922 (w), 1643 (w), 1583 (w), 1491 (w), 1453 (w), 1382 (w), 1331 (w), 1260 (w), 1160 (w), 1089 (w), 1028 (m), 993 (w), 902 (w), 865 (w), 800 (w), 701 (w), 658 (w), 613 (w), 491 (w), 467 (w), 393 (vw). – **MS** (EI, 70 eV), m/z (%): 331/329 (4/3) [M]$^+$, 185/183 (38/7) [M – NMe$_2$ – C$_8$H$_8$]$^+$, 92 (100) [C$_7$H$_8$]$^+$. – **HRMS** (EI, C$_{18}$H$_{20}$N$_1$^{79}Br$_1$) calc. 329.0774, found 329.0772.

(*rac*)-4-Carboxy[2.2]paracyclophane (153)

To a solution of (*rac*)-4-bromo[2.2]paracyclophane (**152**, 2.86 g, 10.0 mol, 1.00 equiv.) in abs. THF (30 mL) was added *n*BuLi (2.5 M in hexane, 6.40 mL, 16.0 mmol, 1.60 equiv.) dropwise at –78 °C. After stirring at –78 °C for 2 h, CO$_2$ was bubbled through the solution *via* a long needle with stirring. The reaction mixture was quenched with water and extracted with 1 M NaOH solution (2 ×20 mL) until all the gas was consumed. The water phases were combined, washed with Et$_2$O (50 mL) and acidified with 6 M HCl until the solution showed acid. The precipitate was filtrated, washed with water and Et$_2$O. The product **153** was obtained after drying under high vacuum as white powder, 1.30 g, 5.16 mmol, 52%.

R$_f$ = 0.12 (*c*-Hex/EtOAc = 9:1). – **¹H NMR** (500 MHz, CD$_2$Cl$_2$) δ/ppm = 11.44 (brs, 1H, COO*H*), 7.20 (d, *J* = 1.9 Hz, 1H, C*H*^{Ar}), 6.66 (dd, *J* = 7.8, 2.0 Hz, 1H, C*H*^{Ar}), 6.55–6.49 (m, 3H, 3 × C*H*^{Ar}), 6.47–6.39 (m, 2H, 2 × C*H*^{Ar}), 4.14–4.05 (m, 1H, C*H*^{PC}), 3.17–3.92 (m, 6H, 6 × C*H*^{PC}), 2.83 (ddd, *J* = 12.9, 10.1, 7.1 Hz, 1H, C*H*^{PC}). – **¹³C NMR** (126 MHz, CD$_2$Cl$_2$) δ/ppm = 172.13 (C$_q$, *C*OOH), 144.24 (C$_q$, *C*^{Ar}), 140.82 (C$_q$, *C*^{Ar}), 140.55 (C$_q$, *C*^{Ar}), 140.14 (C$_q$, *C*^{Ar}), 137.93 (+, CH, *C*^{Ar}), 136.91 (+, CH, *C*^{Ar}), 136.52 (+, CH, *C*^{Ar}), 133.48 (+, CH, *C*^{Ar}), 133.30 (+, CH, *C*^{Ar}), 132.73 (+, CH, *C*^{Ar}), 132.12 (+, CH, *C*^{Ar}), 130.00 (C$_q$, *C*^{Ar}-COOH), 36.67 (–, CH$_2$, *C*^{PC}), 35.63 (–, CH$_2$, *C*^{PC}), 35.48 (–, CH$_2$, *C*^{PC}), 35.31 (–, CH$_2$,

C^{PC}). – **IR** (ATR): $\tilde{v}$/cm^{-1} = 2923 (w), 2852 (w), 1681 (m), 1592, 1555 (w), 1497 (w), 1435 (w), 1418 (w), 1303 (m), 1274 (m), 1200 (w), 1077 (w), 946 (w), 913 (w), 884 (w), 868 (w), 798 (w), 763 (w), 721 (w), 706 (w), 667 (w), 634 (w), 577 (w), 559 (w), 511 (m), 441 (w), 424 (vw). – **MS** (EI, 70 eV), m/z (%): 252 (27) [M]$^+$, 148 (18) [M – C$_8$H$_8$]$^+$, 104 (100) [C$_8$H$_8$]$^+$. – **HRMS** (EI, C$_{17}$H$_{16}$O$_2$) calc. 252.1150, found 252.1149. The analytical data match those reported in the literature.[176,234–236]

(*rac*)-4-Amino[2.2]paracyclophane (161a)

To a solution of (*rac*)-4-bromo[2.2]paracyclophane (**152**, 2.00 g, 7.00 mmol, 1.00 equiv.) in 32 mL of abs. THF was added *n*BuLi (2.5 M in hexane, 3.24 mL, 8.05 mmol, 1.15 equiv.) dropwise in 30 min at –78 °C. After stirring at –78 °C for 1 h, the reaction was allowed to warm to room temperature. A solution of *p*-ABSA (1.93 g, 8.05 mmol, 1.15 equiv.) in 10 mL of THF was added dropwise over 45 min. After stirring overnight, a solution of saturated NH$_4$Cl was added, the water phase was extracted with Et$_2$O (2 × 50 mL), and the combined organic phase was washed with brine, dried over Na$_2$SO$_4$, filtrated and concentrated under vacuum. The resulting crude was used directly in the next step without further purification.

To a solution of above mentioned crude in 20 mL of abs. THF was added NaBH$_4$ (2.65 g, 70.0 mmol, 10.0 equiv.) in portions. 2 mL of abs. MeOH was added slowly over 5 h under reflux, after the reaction was cooled to room temperature, saturated NaHCO$_3$ solution was added, the water phase was extracted with CH$_2$Cl$_2$ (3 × 50 mL), the combined organic phase was washed with brine, dried over Na$_2$SO$_4$, filtrated and concentrated under vacuum. The crude was purified *via* column chromatography (*c*-Hex/EtOAc = 9:1 → 6:1) to give product **161a** as colorless solid, 738 mg, 3.30 mmol, 47%.

R_f = 0.14 (*c*-Hex/EtOAc = 8:1). – **^{1}H NMR** (500 MHz, CDCl$_3$) δ/ppm = 7.21 (dd, *J* = 7.8, 1.9 Hz, 1H, C*H*^{Ar}), 6.64 (dd, *J* = 7.9, 1.9 Hz, 1H, C*H*^{Ar}), 6.44 (ddd, *J* = 7.8, 3.9, 1.9 Hz, 2H, 2 × C*H*^{Ar}), 6.31 (d, *J* = 7.6 Hz, 1H, C*H*^{Ar}), 6.18 (dd, *J* = 7.7, 1.8 Hz, 1H, C*H*^{Ar}), 5.40 (d, *J* = 1.7 Hz, 1H, C*H*^{Ar}), 3.40 (brs, 2H, C^{Ar}-N*H*$_2$), 3.18–2.95 (m, 6H, 6 × C*H*^{PC}), 2.91– 2.82 (m, 1H, C*H*^{PC}), 2.75–2.65 (m, 1H, C*H*^{PC}). – **^{13}C NMR** (126 MHz, CDCl$_3$) δ/ppm = 144.99 (C$_q$, C^{Ar}-NH$_2$), 141.08 (C$_q$, C^{Ar}), 139.00 (C$_q$, C^{Ar}), 138.94 (C$_q$, C^{Ar}), 135.29 (+, CH, C^{Ar}), 133.49 (+, CH, C^{Ar}), 132.47 (–, CH, C^{Ar}), 131.49 (+, CH, C^{Ar}), 126.81 (+, CH, C^{Ar}), 124.48 (C$_q$, C^{Ar}), 122.82 (+, CH, C^{Ar}), 122.29 (+, CH, C^{Ar}), 35.40 (–, CH$_2$, C^{PC}), 34.98 (–, CH$_2$, C^{PC}), 33.02 (–, CH$_2$, C^{PC}), 32.28 (–, CH$_2$, C^{PC}). – **IR** (ATR): $\tilde{v}$/cm^{-1} = 3383 (vw), 2920 (vw), 2848 (vw), 1611 (w), 1497 (w), 1423 (w), 1284 (w), 862 (vw), 794 (w), 715 (w), 658 (w), 509 (w), 490 (w). – **MS** (EI, 70 eV), m/z (%): 223 (30) [M]$^+$, 119 (100) [M – C$_8$H$_8$]$^+$, 104 (8) [C$_8$H$_8$]$^+$, 91 (16) [C$_7$H$_7$]$^+$. – **HRMS** (EI, C$_{16}$H$_{17}$N$_1$)

calc. 223.1361, found 223.1363. The analytical data match those reported in the literature.[173,231,237–240]

(Sₚ,S)/(Rₚ,S)-4-(4'-Phenyloxazolin-2'-yl)[2.2]paracyclophane/*(Sₚ,S)/(Rₚ,S)*-155

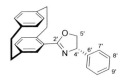

Thionyl chloride (10 mL) was added to (*rac*)-4-carboxyl [2.2]paracyclophane (**153**, 0.650 g, 2.58 mmol, 1.00 equiv.) and the resulting mixture was heated to 60 °C for 4 h. The excess thionyl chloride was removed by evaporation and the final traces were washed with toluene (2 × 5 mL) and removed under vacuum. The resulting crude acetyl chloride was dissolved in 10 mL of CH₂Cl₂ and cooled to 0 °C. A solution of (*S*)-(+)-phenylglycinol (**145c**, 0.710 g, 5.16 mmol, 2.00 equiv.) and Et₃N (0.720 mL, 0.520 g, 5.16 mmol, 2.00 equiv.) in 5 mL of CH₂Cl₂ was added and the reaction mixture was allowed to warm to room temperature. After stirring at room temperature for 24 h, 10 mL of CH₂Cl₂ was added and washed with aq. NaHCO₃ solution (3.5% w/v, 2 × 10 mL) and brine (20 mL). The organic phase was dried over Na₂SO₄, filtered, concentrated and dried under vacuum to give the crude amide as light brown solid.

To a solution of the above mentioned crude and PPh₃ (1.19 g, 4.52 mmol, 1.75 equiv.) in 15 mL of dry MeCN was added CCl₄ (0.440 mL, 0.700 g, 4.52 mmol, 1.75 equiv.) and Et₃N (0.630 mL, 0.460 g, 4.52 mmol, 1.75 equiv.) under argon atmosphere. The mixture was stirred at room temperature overnight. After removing the solvent under reduced pressure, the resulting crude was dissolved in CH₂Cl₂, washed with brine (2 × 20 mL), dried over Na₂SO₄, filtered and concentrated under vacuum. The crude was purified *via* column chromatography (*c*-Hex/EtOAc = 95:5) to obtain the title product *(Sₚ,S)/(Rₚ,S)*-**155** as colorless solid, 549 mg, 1.56 mmol, 60%.

The product was obtained as an inseparable mixture of diastereomers with a ratio of 1:1. An unambiguous assignment of the signals to a definite diastereomer was not possible even with 2D NMR experiments. Therefore, the signals for the diastereomeric mixture are given, same signals are stated with "and" if clearly distinguishable.

R_f = 0.40 (*c*-Hex/EtOAc = 4:1). – **¹H NMR** (500 MHz, CDCl₃) δ/ppm = 7.44–7.40 (m, 2 × 4H, C*H*⁷' and C*H*⁸'), 7.37–7.31 (m, 2 × 1H, C*H*⁹'), 7.19 and 7.18 each (d, *J* = 1.9 Hz, 1H, C*H*ᴬʳ), 6.69–6.62 (m, 2 × 2H, C*H*ᴬʳ), 6.60–6.49 (m, 2 × 4H, C*H*ᴬʳ), 5.47 (ddd, *J* = 15.0, 10.1, 8.3 Hz, 1H, 2 × 1H, C*H*⁴'), 4.80 (ddd, *J* = 14.7, 10.1, 8.3 Hz, 2 × 1H, C*H*⁵'), 4.44–4.12 (m, 2 × 2H, C*H*ᴾᶜ + C*H*⁵'), 3.25–2.99 (m, 2 × 6H, C*H*ᴾᶜ), 2.96–2.84 (m, 2 × 1H, C*H*ᴾᶜ). – **¹³C NMR** (126 MHz, CDCl₃) δ/ppm = 165.29 and 165.04 each (C_q, *C*²') 142.85

and 142.75 each (C_q, $C^{6'}$), 141.40 and 141.38 each (C_q, 2C, C^{Ar}), 139.99 (C_q, 2 × C^{Ar}), 139.80 (C_q, 2 × C^{Ar}), 139.53 (C_q, 2 × C^{Ar}), 136.11 and 136.02 each (+, CH, C^{Ar}), 135.34 and 135.28 each (+, CH, C^{Ar}), 134.66 (+,CH, 2 × C^{Ar}), 133.11 (+,CH, 2 × C^{Ar}), 132.91 and 132.86 each (+, CH, C^{Ar}), 132.48 and 132.42 each (+, CH, C^{Ar}), 131.49 and 131.39 each (+, CH, C^{Ar}), 128.91 and 128.90 each (+, CH, 2 × $C^{8'}$), 127.70 (+, CH, 2 × $C^{9'}$), 126.86 and 127.84 each (+, CH, 2 × $C^{7'}$), 74.40 and 74.13 each (–, CH_2, $C^{5'}$), 70.48 and 70.35 each (+, CH, $C^{4'}$), 36.13 and 36.00 each (–, CH_2, C^{PC}), 35.42 (–, CH_2, 2 × C^{PC}), 35.18 (–, CH_2, 2 × C^{PC}), 35.0 and 35.08 each (–, CH_2, C^{PC}). – **IR** (ATR): $\tilde{v}$/cm^{-1} = 2923 (w), 2850 (vw), 1626 (w), 1588 (w), 1493 (w), 1454 (w), 1343 (w), 1297 (w), 1273 (w), 1247 (vw), 1170 (w), 1129 (vw), 1049 (w), 1027 (w), 975 (w), 954 (w), 911 (w), 885 (w), 836 (vw), 766 (w), 696 (w), 636 (w), 601 (vw), 534 (w), 513 (w), 382 (vw). – **MS** (EI, 70 eV), m/z (%): 353 (87) [M]$^+$, 249 (100) [M – C_8H_8]$^+$, 158 (56), 104 (64) [C_8H_8]$^+$. – **HRMS** (EI, $C_{25}H_{23}O_1N_1$) calc. 353.1780, found 353.1780. The analytical data match those reported in the literautre.[231,241]

(S$_p$,S)/(R$_p$,S)-(Acetonitrile-κN)-(η6-p-cymene)[4-((S)-4'-phenyloxazoline-2'-yl-κN) [2.2]para-cyclophanido-κC5]ruthenium(II)hexafluorophosphate / (S$_p$,S)/(R$_p$,S)-151

(S_p,S)/(R_p,S)-4-(4'-phenyloxazolin-2'-yl)[2.2]paracyclophane ((S_p,S)/(R_p,S)-**155**, 70.6 mg, 0.200 mmol, 1.00 equiv.), dichloro(p-cymene)ruthenium (II) dimer [RuCl$_2$(p-cymene)]$_2$ (50.2 mg, 0.100 mmol, 0.500 equiv.), potassium acetate (29.4 mg, 0.300 mmol, 1.50 equiv.) and potassium hexafluorophosphate (73.6 mg, 0.400 mmol, 2.00 equiv.) were added to an oven dried Schlenk tube. After abs. CH$_3$CN (10 mL) was added, the mixture was heated to 80 °C and stirred under argon atmosphere for 3 days. The reaction mixture was cooled to room temperature and concentrated under reduced pressure. The resulting mixture was purified *via* column chromatography (c-Hex/EtOAc = 1:1 → CH$_3$CN), the product was obtained as a yellow-greenish solid, 90.2 mg, 0.108 mmol, 54%.

As the ligand was obtained as a inseparable mixture of two diastereomers with a ratio of 1:1, the product was therefore obtained as mixture of diastereomers, which could not be separated. An unambiguous assignment of the signals to a definite diastereomer was not possible even with 2D NMR experiments. Therefore, the signals for the diastereomeric mixture are given, the same signals are given with "and" if clearly distinguishable.

R_f = 0.53 (CH$_2$Cl$_2$/CH$_3$CN = 10:1). – **^{1}H NMR** (400 MHz, CD$_3$CN) δ/ppm = 7.62–7.55 (m, 4H, CH^{Ar}), 7.59–7.51 (m, 5H, CH^{Ar}), 7.51–7.37 (m, 1H, CH^{Ar}) 6.73 (dtd, J = 10.0, 7.8, 2.2 Hz, 4H, CH^{Ar}), 6.44 (t, J = 7.4 Hz, 2H, CH^{Ar}), 6.20 (d, J = 6.0 Hz, 1H, C$H^{3''or4''}$) 6.18 (d, J = 6.0 Hz, 1H, C$H^{3''or4''}$), 6.16 (d, J = 5.9 Hz, 1H, C$H^{3''or4''}$), 6.08 (d, J = 7.5 Hz, 2H, CH^{Ar}), 5.94 (d, J = 5.9 Hz, 1H, C$H^{3''or4''}$), 5.75 (d, J = 6.1 Hz, 1H, C$H^{3''or4''}$), 5.68 (d, J = 6.0 Hz, 1H, C$H^{3''or4''}$), 5.31 (td, J = 9.7, 9.3, 8.0 Hz, 2H, CH^{Ar}), 5.26–5.22 (m, 2H, C$H^{4'}$), 5.19 (t, J = 7.9 Hz, 2H, C$H^{5'}$), 5.18 (d, J = 6.0 Hz, 1H, C$H^{3''or4''}$), 5.10 (d, J = 6.1 Hz, 1H, C$H^{3''or4''}$), 4.92 (t, J = 8.2, 1H, C$H^{4'}$) and 4.81 (t, J = 8.2 Hz, 1H, C$H^{4'}$), 3.81 (ddd, J = 13.1, 7.8, 2.6 Hz, 2H, CH^{pC}), 3.62 (dd, J = 13.5, 9.1 Hz, 1H, CH^{pC}), 3.57–3.46 (m, 1H, CH^{pC}), 3.24 (dd, J = 13.1, 9.2 Hz, 1H, CH^{pC}), 3.19–3.07 (m, 3H, CH^{pC}), 3.06–2.70 (m, 8H, CH^{pC}), 2.01 (h, J = 6.9 Hz, 2H, C$H^{6''}$), 0.80 (s, 3H, C$H^{7''}$), 0.72 (d, J = 6.9 Hz, 3H, C$H^{7''}$), 0.63 (d, J = 7.0 Hz, 3H, C$H^{7''}$), 0.48 (brs, 3H, C$H^{7''}$). – **^{13}C NMR** (126 MHz, CD$_3$CN) δ/ppm = 180.11 and 179.69 each (C$_q$, C^{Ar}-Ru), 175.66 and 175.20 each (C$_q$, $C^{2'}$), 150.69 and 150.50 each (C$_q$, C^{Ar}), 141.33 and 141.24 each (C$_q$, C^{Ar}), 140.96 and 140.91 each (C$_q$, C^{Ar}), 140.81 and 140.79 each (C$_q$, C^{Ar}), 140.19 and 140.16 each (C$_q$, C^{Ar}), 139.49 and 139.34 each (C$_q$, C^{Ar}), 134.55 and 134.46 each (+, CH, C^{Ar}), 133.42 and 133.35 each (+, CH, C^{Ar}), 133.19 and 133.17 each (+, CH, C^{Ar}), 133.14 and 133.06 each (+, CH, C^{Ar}), 132.52 and 132.43 each (+, CH, C^{Ar}), 132.37 and 131.98 each (+, CH, C^{Ar}), 130.28 and 130.26 each (+, CH, C^{Ar}), 130.18 and 130.05 each (+, CH, C^{Ar}), 129.83 and 129.78 each (+, CH, 2 × C^{Ar}), 129.73 and 129.65 each (+, CH, 2 × C^{Ar}), 110.24 (C$_q$, 2C, $C^{2''or5''}$), 102.32 (C$_q$, 2C, $C^{2''or5''}$), 93.59 and 93.53 each (+, CH, $C^{3''or4''}$), 91.94 and 90.43 each (+, CH, $C^{3''or4''}$), 84.46 and 83.40 each (+, CH, $C^{3''or4''}$), 82.36 and 82.32 each (+, CH, $C^{3''or4''}$), 79.04 and 78.98 each (–, CH$_2$, $C^{5'}$), 70.34 and 69.49 each (+, CH, $C^{4'}$), 39.33 and 38.69 each (–, CH$_2$, C^{pC}), 36.20 and 36.15 each (–, CH$_2$, C^{pC}), 35.82 and 35.70 each (–, CH$_2$, C^{pC}), 33.91 and 33.34 each (–, CH$_2$, C^{pC}), 31.45 and 31.30 each (+, CH, $C^{6''}$), 22.00 and 21.52 each (+, CH$_3$, 2 × $C^{7''}$), 18.20 and 17.55 each (+, CH$_3$, $C^{1''}$). – **MS** (FAB, 3-NBA), m/z (%): 588 (100) [M – PF$_6$ – CH$_3$CN]$^+$, 454 [M – PF$_6$–CH$_3$CN–cymene]$^+$, – **HRMS** (FAB, C$_{35}$H$_{36}$O$_1$N$_1$^{102}Ru$_1$, [M+H]$^+$): calc. 588.1840, found 588.1842.

6.2.3 Synthesis and Characterization of the Saturated α-Diazocarbonyl Compounds

Phenyl 2-bromopropanoate (115)

To a solution of phenol (4.00 g, 42.5 mmol, 1.00 equiv.) and pyridine (6.70 g, 85.0 mmol, 2.00 equiv.) in 40 mL of CH_3CN was added 2-bromoisopropionyl bromide (**114**, 18.3 g, 85.0 mmol, 2.00 equiv.) at 0 °C. After stirring at 0 °C for 30 min, the reaction mixture was quenched with 30 mL of H_2O and extracted with CH_2Cl_2 (3 × 50 mL). The combined organic layers were washed with H_2O (3 × 50 mL) and dried over Na_2SO_4. After removal of the solvent by evaporation, the resulting crude material was purified *via* column chromatography (*c*-Hex/EtOAc = 8:1) to give **115** as a colorless oil, 9.40 g, 39.2 mmol, 97%.

R_f = 0.50 (*c*-Hex/EtOAc = 10:1). – ^{1}**H NMR** (300 MHz, CDCl$_3$) δ/ppm = 7.43 (t, J = 7.7 Hz, 2H, CH^{Ar}), 7.28 (t, J = 7.7 Hz, 1H, CH^{Ar}), 7.16 (d, J = 8.0 Hz, 1H, CH^{Ar}), 4.26 (q, J = 7.0 Hz, 1H, C*H*Br), 1.98 (d, J = 7.6 Hz, 3H, CH_3). – ^{13}**C NMR** (75 MHz, CDCl$_3$) δ/ppm = 169.28 (C$_q$, *C*=O), 150.99 (C$_q$, C^{Ar}), 130.08 (+, CH, 2 × C^{Ar}), 126.85 (+, CH, C^{Ar}), 121.47 (+, CH, 2 × C^{Ar}), 40.23 (+, CH, *C*HBr), 22.00 (+, *C*H$_3$). The analytical data match those reported in literature.[128]

Phenyl 2-diazopropanoate (116)

Phenyl 2-bromopropionate (**115**, 1.60 g, 7.00 mmol, 1.00 equiv.) and *N,N'*-ditosylhydrazine (4.80 g, 14.0 mmol, 2.00 equiv.) were dissolved in THF (20 mL). The solution was cooled to 0 °C in ice bath. After DBU (5.20 mL, 5.33 g, 35.0 mmol, 5.00 equiv.) was added dropwise, the mixture was stirred at 0 °C for 30 min. The reaction was quenched with sat. NaHCO$_3$ (25 mL). The resulting mixture was extracted with CH_2Cl_2 (30 mL × 3). The combined organic layer was washed with H_2O (30 mL × 3) and dried over Na_2SO_4. After the solvent was removed by evaporation, the resulting crude material was purified by column chromatography (pentane/Et$_2$O = 15:1) to give **116** as a yellow liquid, 0.26 g, 1.47 mmol, 21%.

R_f = 0.40 (*c*-Hex/EtOAc = 10:1) – ^{1}**H NMR** (300 MHz, CDCl$_3$) δ/ppm = 7.43–7.34 (m, 2H, Ph), 7.29–7.20 (m, 1H, Ph), 7.18–7.02 (m, 2H, Ph), 2.09 (s, 3H, CH_3N$_2$). – ^{13}**C NMR** (75 MHz, CDCl$_3$) δ/ppm = 150.62 (C$_q$, *C*O$_2$Ph), 136.48 (C$_q$, C^{Ar}), 129.32 (+, CH, 2 × C^{Ar}), 125.60 (+, CH, C^{Ar}), 121.67 (+, CH, 2 × C^{Ar}), 1.02 (+, CH$_3$, *C*H$_3$N$_2$). The analytical data match those reported in literature.[128]

Methyl 2-diazo-2-phenylacetate (118)

 p-Toluenesulfonyl hydrazide (1.86 g, 10.0 mmol, 1.00 equiv.) was added to a solution of methyl 2-oxo-2-phenylacetate (**117**, 1.64 g, 10.0 mmol, 1.00 equiv.) in toluene (50 mL). The resulting mixture was heated under reflux with azeotropic removal of water for 9 h and then concentrated *in vacuo*. The residue was dissolved in CH$_2$Cl$_2$ (30 mL) and Et$_3$N (1.83 mL,1.33 g, 10.0 mmol, 1.00 equiv.) was slowly added under argon atmosphere. After stirring at 25 °C for 36 h, the resulting mixture was washed with 20 mL of water and 20 mL of brine, dried over Na$_2$SO$_4$, filtered, and concentrated *in vacuo*. The residue was purified by flash chromatography on silica gel (*c*-Hex/EtOAc = 5:1) to provide **118** as a yellow liquid, 1.39 g, 7.90 mmol, 79%.

R_f = 0.37 (*c*-Hex/EtOAc = 10:1) – **¹H NMR** (300 MHz, CDCl$_3$) δ/ppm = 7.45–7.35 (m, 2H, 2 × C*H*^{Ar}), 7.27 (dd, *J* = 8.6, 7.1 Hz, 2H, 2 × C*H*^{Ar}), 7.12–7.02 (m, 1H, C*H*^{Ar}), 3.75 (s, 3H, OC*H*$_3$). – **¹³C NMR** (75 MHz, CDCl$_3$) δ/ppm = 155.94 (C$_q$, *C*=O), 128.89 (C$_q$, *C*^{Ar}), 125.64 (+, CH, *C*^{Ar}), 125.49 (+, CH, 2 × *C*^{Ar}), 123.98 (+, CH, 2 × *C*^{Ar}), 51.96 (+, CH$_3$, OMe). The analytical data match those reported in literature.[56]

Ethyl 2-methyl-3-oxobutanoate (112a)

General procedure 6 (GP 6) was followed by adding ethyl acetoacetate (4.86 mL, 5.00 g, 38.4 mmol, 1.00 equiv.) to a suspension of NaH (60% in oil, 1.54 g, 38.4 mmol, 1.00 equiv.) in 20 mL of dry THF. Iodomethane (2.90 mL, 6.54 g, 46.1 mmol, 1.20 equiv.) was added dropwise and the mixture was stirred under reflux overnight. The product **112a** was obtained *via* column chromatography (*c*-Hex/EtOAc = 15:1) as a colorless liquid, 2.27 g, 15.7 mmol, 41%.

R_f = 0.30 (*c*-Hex/EtOAc = 8:1). – **¹H NMR** (300 MHz, CDCl$_3$) δ/ppm = 4.20 (q, 2H, *J* = 7.2 Hz, OC*H*$_2$CH$_3$), 3.51 (q, 1H, *J* = 7.1 Hz, C*H*CH$_3$), 2.25 (s, 3H, COC*H*$_3$), 1.34 (d, 3H, *J* = 7.2 Hz, CHC*H*$_3$), 1.28 (t, 3H, *J* = 7.2 Hz, OCH$_2$C*H*$_3$). – **¹³C NMR** (75 MHz, CDCl$_3$) δ/ppm = 204.10 (C$_q$, *C*=O), 170.98 (C$_q$, *C*O$_2$CH$_2$CH$_3$), 61.79 (+, CH, *C*HCH$_3$), 54.07 (–, CH$_2$, *C*H$_2$CH$_3$), 28.87 (+, CH$_3$, COC*H*$_3$), 14.52 (+, *C*H$_3$), 13.16. (+, *C*H$_3$). The analytical data match those reported in literature.[242]

tert-Butyl 2-methyl-3-oxobutanoate (112b)

General procedure 6 (GP 6) was followed by adding *tert*-butyl acetoacetate (4.95 mL, 4.75 g, 30.0 mmol, 1.00 equiv.) to a suspension of NaH (60% in oil, 1.20 g, 30.0 mmol, 1.00 equiv.) in 20 mL of dry THF. Iodomethane (1.64 mL, 4.68 g, 33.0 mmol, 1.10 equiv.) was added dropwise and the mixture was stirred under reflux overnight. The product **112b** was obtained *via* column chromatography (*c*-Hex/EtOAc = 15:1) as a colorless liquid, 2.46 g, 14.3 mmol, 48%.

R_f = 0.42 (*c*-Hex/EtOAc = 8:1). – **¹H NMR** (300 MHz, CDCl$_3$) δ/ppm = 3.32 (q, *J* = 7.1 Hz, 1H, C*H*CH$_3$), 2.13 (s, 3H, COC*H*$_3$), 1.37 (s, 9H, C(C*H*$_3$)$_3$), 1.18 (d, *J* = 7.1 Hz, 3H, CHC*H*$_3$). – **¹³C NMR** (75 MHz, CDCl$_3$) δ/ppm = 203.82 (C$_q$, *C*=O), 169.64 (C$_q$, *C*O$_2$tBu), 81.59 (C$_q$, *C*(CH$_3$)$_3$), 54.56 (+, CH, *C*HCH$_3$), 28.29 (+, CH$_3$, CO*C*H$_3$) 27.80 (+, CH$_3$, OC(*C*H$_3$)$_3$), 12.54 (+, CHC*H*$_3$). The analytical data match those reported in literature.[128]

tert-Butyl 2-diazopropanoate (113b)

General procedure 7 (GP 7) was followed by adding Et$_3$N (4.50 mL, 4.34 g, 42.9 mmol, 3.00 equiv.) to a solution of *tert*-butyl 2-methyl-3-oxobutanoate (**112b**, 2.05 g, 14.3 mmol, 1.00 equiv.) in 15 mL of CH$_3$CN. *p*-ABSA (6.79 g, 28.3 mmol, 2.00 equiv.) in 15 mL of CH$_3$CN was then added dropwise and the mixture was stirred under room temperature overnight. The product **113b** was purified *via* column chromatography (*c*-Hex/EtOAc = 10:1) and isolated as a yellow liquid, 1.03 g, 6.58 mmol, 46%.

R_f = 0.40 (*c*-Hex/EtOAc = 10:1). – **¹H NMR** (300 MHz, CDCl$_3$) δ/ppm = 1.89 (s, 3H, C*H*$_3$N$_2$), 1.45 (s, 9H, C(C*H*$_3$)$_3$). – **¹³C NMR** (75 MHz, CDCl$_3$) δ/ppm = 167.33 (C$_q$, *C*O$_2$tBu), 80.98 (C$_q$, *C*(CH$_3$)$_3$), 28.32 (+, CH$_3$, 3 × *C*H$_3$), 8.37 (+, CH$_3$, *C*H$_3$N$_2$). The analytical data match those reported in literature.[128]

Benzyl 2-methylacetoacetate (112c)

General procedure 6 (GP 6) was followed by adding benzyl acetoacetate (5.00 g, 26.0 mmol, 1.00 equiv.) to a suspension of NaH (60% in oil, 1.04 g, 26.0 mmol, 1.00 equiv.) in 25 mL of dry THF. Iodomethane (1.94 mL, 4.43 g, 31.2 mmol, 1.20 equiv.) was added dropwise and the mixture was stirred under reflux overnight. The product **112c** was obtained *via* column chromatography (*c*-Hex/EtOAc = 10:1) as a colorless liquid, 3.46 g, 16.8 mmol, 65%.

R_f = 0.25 (*c*-Hex/EtOAc = 8:1). – **¹H NMR** (300 MHz, CDCl₃) δ/ppm = 7.94–7.03 (m, 5H, CH₂*Ph*), 5.17 (s, 2H, C*H₂*Ph), 3.55 (q, *J* = 7.1 Hz, 1H, C*H*CH₃), 2.18 (s, 3H, C*H₃*O), 1.35 (d, *J* = 7.2 Hz, 3H, CHC*H₃*). – **¹³C NMR** (75 MHz, CDCl₃) δ/ppm = 203.88 (C_q, *C*=O), 170.80 (C_q, *C*O₂Bn), 135.84 (C_q, *C*^Ar), 128.88 (+, CH, 2 × *C*^Ar), 128.87 (+, CH, *C*^Ar), 128.63 (+, CH, 2 × *C*^Ar), 67.54 (–, CH₂, *C*H₂Ph), 54.20 (+, CH, *C*HCH₃), 28.89 (+, CH₃, *C*H₃O), 13.30 (+, CH₃, CH*C*H₃). The analytical data match those reported in literature.[128]

Benzyl 2-diazopropanoate (113c)

General procedure 7 (GP 7) was followed by adding Et₃N (1.25 mL, 0.910 g, 9.00 mmol, 3.00 equiv.) to a solution of benzyl 2-methylacetoacetone (**112c**, 0.620 g, 3.00 mmol, 1.00 equiv.) in 5 mL of CH₃CN. *p*-ABSA (1.44 g, 6.00 mmol, 2.00 equiv.) in 5 mL of CH₃CN was then added dropwise and the mixture was stirred under room temperature overnight. The product **113c** was purified *via* column chromatography (pentane/Et₂O = 10:1) and isolated as a yellow liquid, 0.540 g, 2.85 mmol, 95%.

R_f = 0.44 (pentane/Et₂O = 8:1). – **¹H NMR** (300 MHz, CDCl₃) δ/ppm = 7.75–7.05 (m, 5H, CH₂*Ph*), 5.21 (s, 2H, C*H₂*Ph), 1.98 (s, 3H, C*H₃*N₂). – **¹³C NMR** (75 MHz, CDCl₃) δ/ppm = 136.16 (C_q, *C*^Ar), 128.50 (+, CH, 2 × *C*^Ar), 128.14 (+, CH, *C*^Ar), 128.00 (+, CH, 2 × *C*^Ar), 66.32 (–, CH₂, *C*H₂Ph), 8.44 (+, CH₃, *C*H₃N₂). The analytical data match those reported in literature.[128]

6.2.4 Synthesis and Characterization of the N–H Insertion Products from the Unsaturated α-Diazocarbonyl Substrates

Methyl 2-phenyl-2-(phenylamino)acetate (170b)

General procedure 4a (GP 4a) was followed by adding 1 mL of abs. CH_2Cl_2 to $(S_p,S)/(R_p,S)$-**144a** (2.60 mg, 6.00 μmol, 6 mol%), NaBArF (5.20 mg, 6.00 μmol, 6 mol%) and $Cu(MeCN)_4PF_6$ (1.6 mg, 5.00 μmol, 5 mol%) in an oven dried vial under argon atmosphere. After stirring at 40 °C for 2 h, methyl 2-diazo-2-phenylacetate (**118**, 17.6 mg, 1.00 mmol, 1.00 equiv.) and aniline (11.2 mg, 1.20 mmol, 1.20 equiv.) were added subsequently and the mixture was stirred under room temperature for 2 h. The product **170b** was obtained *via* flash chromatography (*c*-Hex/EtOAc = 5:1) as colorless solid, 23.6 mg, 0.98 mmol, 98%.

– ^{1}**H NMR** (300 MHz, CDCl$_3$) δ/ppm = 7.42 (d, J = 7.7 Hz, 2H, $2 \times CH^{Ar}$), 7.27 (qd, J = 7.5, 6.4, 2.6 Hz, 3H, $3 \times CH^{Ar}$), 7.04 (t, J = 7.9 Hz, 2H, $2 \times CH^{Ar}$), 6.62 (t, J = 7.3 Hz, 1H, CH^{Ar}), 6.48 (d, J = 7.7 Hz, 2H, $2 \times CH^{Ar}$), 5.01 (d, J = 5.9 Hz, 1H, CHN), 4.88 (s, 1H, NH), 3.65 (s, 3H, OCH_3). – ^{13}**C NMR** (75 MHz, CDCl$_3$) δ/ppm = 172.44 (C$_q$, C=O), 145.85 (C$_q$, C^{Ar}), 137.59 (C$_q$, C^{Ar}), 129.34 (+, CH, $2 \times C^{Ar}$), 128.88 (+, CH, $2 \times C^{Ar}$), 128.32 (+, CH, C^{Ar}), 127.21 (+, CH, $2 \times C^{Ar}$), 117.98 (+, CH, C^{Ar}), 113.32 (+, CH, $2 \times C^{Ar}$), 60.62 (+, CH, CHN), 52.10 (+, CH$_3$, OCH_3). The analytical data match those reported in the literature.[58,243–245]

tert-Butyl phenylalaninate (170j)

General procedure 4a (GP 4a) was followed by adding 1 mL of abs. CH_2Cl_2 to $(S_p,S)/(R_p,S)$-**144a** (2.60 mg, 6.00 μmol, 6 mol%), NaBArF (5.20 mg, 6.00 μmol, 6 mol%) and $Cu(MeCN)_4PF_6$ (1.6 mg, 5.00 μmol, 5 mol%) in an oven dried vial under argon atmosphere. After stirring at 40 °C for 2 h, *tert*-butyl 2-diazopropanoate (**113b**, 15.6 mg, 1.00 mmol, 1.00 equiv.) and aniline (11.2 mg, 1.20 mmol, 1.20 equiv.) were added subsequently and the mixture was stirred under room temperature for 2 h. The product **170j** was obtained *via* flash chromatography (*c*-Hex/EtOAc = 8:1) as a light-yellow liquid, 16.4 mg, 0.74 mmol, 74%.

– ^{1}**H NMR** (300 MHz, CDCl$_3$) δ/ppm = 7.21–7.11 (m, 3H, CH^{Ar}), 6.73 (t, J = 7.3 Hz, 1H, CH^{Ar}), 6.61 (d, J = 7.7 Hz, 2H, CH^{Ar}), 4.02 (q, J = 6.9 Hz, 1H, CHN), 1.44 (s, 9H, C(CH_3)$_3$), 1.43 (d, J = 6.9 Hz, 3H). – ^{13}**C NMR** (75 MHz, CDCl$_3$) δ/ppm = 174.38 (C$_q$, CO_2tBu), 147.36 (C$_q$, C^{Ar}), 129.80 (+, CH, $2 \times C^{Ar}$), 118.62 (+, CH, C^{Ar}), 114.02 (+, CH, $2 \times C^{Ar}$),

82.00 (C$_q$, C(CH$_3$)$_3$), 53.16 (+, CH, CHNH), 28.56 (+, CH$_3$, 3 × CH$_3$), 19.45 (+, CH$_3$, CHCH$_3$). The analytical data match those reported in the literature.[49]

Benzyl o-tolylalaninate (170g)

General procedure 4a (GP 4a) was followed by adding 1 mL of abs. CH$_2$Cl$_2$ to (S_p,S)/(R_p,S)-**144a** (2.60 mg, 6.00 μmol, 6 mol%), NaBArF (5.20 mg, 6.00 μmol, 6 mol%) and Cu(MeCN)$_4$PF$_6$ (1.6 mg, 5.00 μmol, 5 mol%) in an oven dried vial under argon atomosphere. After stirring at 40 °C for 2 h, benzyl 2-diazopropanoate (**113c**, 19.0 mg, 1.00 mmol, 1.00 equiv.) and o-toluidine (12.9 mg, 1.20 mmol, 1.20 equiv.) were added subsequently and the mixture was stirred under room temperature for 2 h. The product **170g** was obtained *via* column chromatography (c-Hex/EtOAc = 5:1) as colorless solid, 18.3 mg, 0.68 mmol, 68%.

– **^{1}H NMR** (300 MHz, CDCl$_3$) δ/ppm = 7.44–7.27 (m, 5H, CH$_2$Ph), 7.15–7.05 (m, 2H, Ph), 6.72 (td, J = 7.4, 1.2 Hz, 1H, Ph), 6.55 (dd, J = 8.4, 1.2 Hz, 1H, Ph), 5.19 (s, 2H, CH_2Ph), 4.27 (q, J = 6.9 Hz, 1H, CHCH$_3$), 4.08 (s, 1H, NH), 2.21 (s, 3H, CH_3), 1.55 (d, J = 6.9 Hz, 3H, CHCH_3). – **^{13}C NMR** (75 MHz, CDCl$_3$) δ/ppm = 174.71 (C$_q$, CO$_2$Bn), 144.74 (C$_q$, C^{Ar}), 135.72 (C$_q$, C^{Ar}), 130.52 (C$_q$, C^{Ar}), 128.70 (+, CH, 2 × C^{Ar}), 128.45 (+, CH, C^{Ar}), 128.23 (+, CH, 2 × C^{Ar}), 127.22 (+, CH, C^{Ar}), 122.80 (+, CH, C^{Ar}), 118.08 (+, CH, C^{Ar}), 110.54 (+, CH, C^{Ar}), 66.97 (–, CH$_2$, CH$_2$Ph), 52.18 (+, CH, CHN), 19.20 (+, CH$_3$), 17.57 (+, CH$_3$).

Benzyl (2-methoxyphenyl)alaninate (170d)

General procedure 4a (GP 4a) was followed by adding 1 mL of abs. CH$_2$Cl$_2$ to (S_p,S)/(R_p,S)-**144a** (2.60 mg, 6.00 μmol, 6 mol%), NaBArF (5.20 mg, 6.00 μmol, 6 mol%) and Cu(MeCN)$_4$PF$_6$ (1.6 mg, 5.00 μmol, 5 mol%) in an oven dried vial under argon atomosphere. After stirring at 40 °C for 2 h, benzyl 2-diazopropanoate (**113c**, 19.0 mg, 1.00 mmol, 1.00 equiv.) and o-anisidine (14.8 mg, 1.20 mmol, 1.20 equiv.) were added subsequently and the mixture was stirred under room temperature for 2 h. The product **170d** was obtained *via* column chromatography (c-Hex/EtOAc = 5:1) as light yellow solid, 23.3 mg, 0.82 mmol, 82%.

– **^{1}H NMR** (300 MHz, CDCl$_3$) δ/ppm = 7.40–7.06 (m, 5H, CH$_2$$Ph$), 6.78–6.66 (m, 2H, 2 × CH^{Ar}), 6.62 (ddd, J = 8.2, 7.2, 1.6 Hz, 1H, CH^{Ar}), 6.44 (dd, J = 7.6, 1.6 Hz, 1H, CH^{Ar}),

5.07 (s, 2H, CH$_2$Ph), 4.64 (s, 1H, N*H*), 4.12 (q, *J* = 7.0 Hz, 1H, C*H*N), 3.76 (s, 3H, OC*H*$_3$), 1.44 (d, *J* = 6.9 Hz, 3H, CHC*H*$_3$). – **^{13}C NMR** (75 MHz, CDCl$_3$) δ/ppm = 174.59 (C$_q$, CO$_2$Bn), 147.22 (C$_q$, *C*^{Ar}), 136.69 (C$_q$, *C*^{Ar}), 135.84 (C$_q$, *C*^{Ar}), 128.65 (+, CH, 2 × *C*^{Ar}), 128.36 (+, CH, *C*^{Ar}), 128.20 (+, CH, 2 × *C*^{Ar}), 121.30 (+, CH, *C*^{Ar}), 117.72 (+, CH, *C*^{Ar}), 110.61 (+, CH, *C*^{Ar}), 109.94 (+, CH, *C*^{Ar}), 66.83 (–, CH$_2$, *C*H$_2$Ph), 55.57 (+, CH$_3$, O*C*H$_3$), 52.03 (+, CH, *C*HN), 18.99 (+, *C*H$_3$). The analytical data match those reported in the literature.[58]

Benzyl *p*-tolylalaninate (170h)

General procedure 4a (GP 4a) was followed by adding 1 mL of abs. CH$_2$Cl$_2$ to (S$_p$,S)/(R$_p$,S)-**144a** (2.60 mg, 6.00 μmol, 6 mol%), NaBArF (5.20 mg, 6.00 μmol, 6 mol%) and Cu(MeCN)$_4$PF$_6$ (1.6 mg, 5.00 μmol, 5 mol%) in an oven dried vial under argon atomosphere. After stirring at 40 °C for 2 h, benzyl 2-diazopropanoate (**113c**, 19.0 mg, 1.00 mmol, 1.00 equiv.) and *p*-toluidine (12.9 mg, 1.20 mmol, 1.20 equiv.) were added subsequently and the mixture was stirred under room temperature for 2 h. The product **170h** was obtained *via* column chromatography (*c*-Hex/EtOAc = 4:1) as light yellow solid, 18.8 mg, 0.70 mmol, 70%.

R$_f$ = 0.31 (*c*-Hex/EtOAc = 5:1). – **^{1}H NMR** (300 MHz, CDCl$_3$) δ/ppm = 7.34–7.14 (m, 5H, CH$_2$*Ph*), 6.89 (d, *J* = 8.1 Hz, 2H, 2 × C*H*^{Ar}), 6.45 (d, *J* = 8.4 Hz, 2H, 2 × C*H*^{Ar}), 5.06 (s, 2H, C*H*$_2$Ph), 4.09 (q, *J* = 7.0 Hz, 1H, C*H*N), 3.93 (s, 1H, N*H*), 2.16 (s, 3H, C*H*$_3$), 1.39 (d, *J* = 6.9 Hz, 3H, CHC*H*$_3$). – **^{13}C NMR** (75 MHz, CDCl$_3$) δ/ppm = 174.76 (C$_q$, CO$_2$Bn), 144.43 (C$_q$, *C*^{Ar}), 135.76 (C$_q$, *C*^{Ar}), 129.95 (C$_q$, *C*^{Ar}), 128.68 (+, CH, 2 × C*H*^{Ar}), 128.41 (+, CH, C*H*^{Ar}), 128.25 (+, CH, 2 × C*H*^{Ar}), 127.82 (+, CH, C*H*^{Ar}), 113.90 (+, CH, C*H*^{Ar}), 66.88 (–, CH$_2$, *C*H$_2$Ph), 52.66 (+, CH, *C*HN), 20.54 (+, CH$_3$), 19.06 (+, CH$_3$).

Benzyl (3-methoxyphenyl)alaninate (170e)

General procedure 4a (GP 4a) was followed by adding 1 mL of abs. CH$_2$Cl$_2$ to (S$_p$,S)/(R$_p$,S)-**144a** (2.60 mg, 6.00 μmol, 6 mol%), NaBArF (5.20 mg, 6.00 μmol, 6 mol%) and Cu(MeCN)$_4$PF$_6$ (1.6 mg, 5.00 μmol, 5 mol%) in an oven dried vial under argon atomosphere. After stirring at 40 °C for 2 h, benzyl 2-diazopropanoate (**113c**, 19.0 mg, 1.00 mmol, 1.00 equiv.) and *m*-anisidine (14.8 mg, 1.20 mmol, 1.20 equiv.) were added subsequently and the mixture was stirred under room temperature for 2 h. The product **170e** was obtained *via* column chromatography (*c*-Hex/EtOAc = 5:1) as light-yellow solid, 15.1 mg, 53%.

– **¹H NMR** (300 MHz, CDCl₃) δ/ppm = 7.39–7.27 (m, 5H, CH₂*Ph*), 7.07 (t, *J* = 8.1 Hz, 1H, C*H*^{Ar}), 6.32 (dd, *J* = 8.2, 2.3 Hz, 1H, C*H*^{Ar}), 6.25–6.19 (m, 1H, C*H*^{Ar}), 6.16 (t, *J* = 2.3 Hz, 1H, C*H*^{Ar}), 5.16 (s, 2H, C*H*₂Ph), 4.19 (q, *J* = 6.9 Hz, 1H, C*H*N), 3.74 (s, 1H, N*H*), 1.48 (d, *J* = 6.9 Hz, 3H, CHC*H*₃). – **¹³C NMR** (75 MHz, CDCl₃) δ/ppm = 174.37 (C$_q$, *C*O₂Bn), 160.84 (C$_q$, *C*^{Ar}), 147.93 (C$_q$, *C*^{Ar}), 135.56 (C$_q$, *C*^{Ar}), 130.11 (+, CH, *C*^{Ar}), 128.57 (+, CH, 2 × *C*^{Ar}), 128.32 (+, CH, *C*^{Ar}), 128.15 (+, CH, 2 × *C*^{Ar}), 106.37 (+, CH, *C*^{Ar}), 103.73 (+, CH, *C*^{Ar}), 99.50 (+, CH, *C*^{Ar}), 66.85 (–, CH₂, *C*H₂Ph), 55.05 (+, CH₃, O*C*H₃), 52.08 (+, CH, *C*HN), 18.87 (+, *C*H₃).

Phenyl phenylalaninate (170i)

General procedure 4a (GP 4a) was followed by adding 1 mL of abs. CH₂Cl₂ to (*S*$_p$,*S*)/(*R*$_p$,*S*)-**144a** (2.60 mg, 6.00 μmol, 6 mol%), NaBArF (5.20 mg, 6.00 μmol, 6 mol%) and Cu(MeCN)₄PF₆ (1.6 mg, 5.00 μmol, 5 mol%) in an oven dried vial under argon atomosphere. After stirring at 40 °C for 2 h, phenyl 2-diazopropanoate (**118**, 17.6 mg, 1.00 mmol, 1.00 equiv.) and aniline (11.2 mg, 1.20 mmol, 1.20 equiv.) were added subsequently and the mixture was stirred under room temperature for 2 h. The product **170i** was obtained *via* column chromatography (*c*-Hex/EtOAc = 5:1) as light-yellow liquid, 16.3 mg, 0.68 mmol, 68%.

– **¹H NMR** (300 MHz, CDCl₃) δ/ppm = 7.29 (td, *J* = 7.4, 6.8, 1.3 Hz, 2H, 2 × C*H*^{Ar}), 7.20–7.03 (m, 3H, 3 × C*H*^{Ar}), 6.98–6.87 (m, 2H, 2 × C*H*^{Ar}), 6.72 (tt, *J* = 7.3, 1.1 Hz, 1H, C*H*^{Ar}), 6.64 (dt, *J* = 7.7, 1.1 Hz, 2H, 2 × C*H*^{Ar}), 4.32 (q, *J* = 7.0 Hz, 1H, C*H*N), 4.12 (s, 1H, N*H*), 1.58 (d, *J* = 6.9 Hz, 3H, CHC*H*₃).

Benzyl phenylalaninate (170c)

General procedure 4a (GP 4a) was followed by adding 1 mL of abs. CH₂Cl₂ to (*S*$_p$,*S*)/(*R*$_p$,*S*)-**144a** (2.60 mg, 6.00 μmol, 6 mol%), NaBArF (5.20 mg, 6.00 μmol, 6 mol%) and Cu(MeCN)₄PF₆ (1.6 mg, 5.00 μmol, 5 mol%) in an oven dried vial under argon atomosphere. After stirring at 40 °C for 2 h, benzyl 2-diazopropanoate (**113c**, 19.0 mg, 1.00 mmol, 1.00 equiv.) and aniline (11.2 mg, 1.20 mmol, 1.20 equiv.) were added subsequently and the mixture was stirred under room temperature for 2 h. The product **170c** was obtained as light yellow solid (c-Hex/EtOAc = 4:1), 23.9 mg, 0.94 mmol, 94%.

R_f = 0.33 (c-Hex/EtOAc = 5:1). – **^{1}H NMR** (300 MHz, CDCl$_3$) δ/ppm = 7.31–7.16 (m, 5H, CH$_2$P*h*), 7.15–7.02 (m, 2H, 2 × C*H*^{Ar}), 6.73 (tt, J = 7.3, 1.1 Hz, 1H, C*H*^{Ar}), 6.63 (dd, J = 8.6, 1.2 Hz, 2H, 2 × C*H*^{Ar}), 5.07 (s, 2H, C*H*$_2$Ph), 4.14 (q, J = 7.0 Hz, 1H, C*H*N), 3.98 (s, 1H, N*H*), 1.42 (d, J = 7.0 Hz, 3H, CHC*H*$_3$). – **^{13}C NMR** (75 MHz, CDCl$_3$) δ/ppm = 174.52 (C$_q$, *C*O$_2$Bn), 146.60 (C$_q$, *C*^{Ar}), 135.64 (C$_q$, *C*^{Ar}), 129.51 (+, CH, 2 × *C*^{Ar}), 128.58 (+, CH, 2 × *C*^{Ar}), 128.33 (+, CH, *C*^{Ar}), 128.12 (+, CH, 2 × *C*^{Ar}); 117.98 (+, CH, *C*^{Ar}), 119.12 (+, CH, 2 × *C*^{Ar}), 66.79 (–, CH$_2$, *C*H$_2$Ph), 52.11 (+, CH, *C*HN), 18.78 (+, *C*H$_3$). The analytical data match those reported in the literature.[246]

Benzyl (4-methoxyphenyl)alaninate (170f)

General procedure 4a (GP 4a) was followed by adding 1 mL of abs. CH$_2$Cl$_2$ to $(S_p,S)/(R_p,S)$-**144a** (2.60 mg, 6.00 μmol, 6 mol%), NaBArF (5.20 mg, 6.00 μmol, 6 mol%) and Cu(MeCN)$_4$PF$_6$ (1.6 mg, 5.00 μmol, 5 mol%) in an oven dried vial under argon atomosphere. After stirring at 40 °C for 2 h, benzyl 2-diazopropanoate (**113c**, 19.0 mg, 1.00 mmol, 1.00 equiv.) and p-anisidine (14.8 mg, 1.20 mmol, 1.20 equiv.) were added subsequently and the mixture was stirred under room temperature for 2 h. The product **170f** was obtained via column chromatography (c-Hex/EtOAc = 4:1) as light-yellow solid, 20.0 mg, 0.70 mmol, 70%.

– **^{1}H NMR** (300 MHz, CDCl$_3$) δ/ppm = 7.38–7.03 (m, 5H, CH$_2$P*h*), 6.72–6.61 (m, 2H, 2 × C*H*^{Ar}), 6.51 (d, J = 9.0 Hz, 2H, 2 × C*H*^{Ar}), 5.06 (s, 2H, C*H*$_2$Ph), 4.04 (q, J = 6.9 Hz, 1H, C*H*N), 3.66 (s, 3H, C*H*$_3$). – **^{13}C NMR** (75 MHz, CDCl$_3$) δ/ppm = 175.30 (C$_q$, CO$_2$Bn), 153.40 (C$_q$, *C*^{Ar}), 141.26 89 (C$_q$, *C*^{Ar}), 136.17 (C$_q$, *C*^{Ar}), 129.10 (+, CH, 2 × *C*^{Ar}), 128.85 (+, CH, *C*^{Ar}), 128.68 (+, CH, 2 × *C*^{Ar}), 115.75 (+, CH, 2 × *C*^{Ar}), 115.46 (+, CH, 2 × *C*^{Ar}), 67.29 (–, CH$_2$, *C*H$_2$Ph), 56.26 (+, CH$_3$ O*C*H$_3$,), 53.83 (+, CH, *C*HN), 19.54 (+, *C*H$_3$).

6.2.5 Synthesis and Characterization of the Substituted Buta-2,3-Dienoates

Methyl buta-2, 3-dienoate (143a)

General procedure 1 (GP 1) was followed by adding Et$_3$N (6.42 mL, 5.07 g, 50.0 mmol, 1.00 equiv.) to a solution of ylide **142a** (17.2 g, 50.0 mmol, 1.00 equiv.) in 150 mL of CH$_2$Cl$_2$, acetyl chloride (4.32 g, 55.0 mmol, 1.10 equiv.) was added dropwise and the reaction was stirred under room temperature for 6 hours. Product **143a** was isolated *via* column chromatography (pentane/Et$_2$O = 3:1) as a colorless liquid, 4.40 g, 44.0 mmol, 88%.

R_f = 0.54 (pentane/Et$_2$O = 4:1). – **^{1}H NMR** (300 MHz, CDCl$_3$) δ/ppm = 5.63 (d, ^4J = 6.5 Hz, 1H, C*H*CO$_2$Me), 5.21 (d, $^2J_{gem}$ = 6.5 Hz, 2H, C*H$_2$*=), 3.73 (s, 3H, OC*H$_3$*). – **^{13}C NMR** (75 MHz, CDCl$_3$) δ/ppm = 215.82 (C$_q$, *C*=CH), 166.13 (C$_q$, *C*O$_2$Me), 87.69 (+, CH, *C*=CH), 79.3 (–, CH$_2$, *C*H$_2$=C), 52.07 (+, CH$_3$, O*C*H$_3$). – **IR** (ATR): ṽ/cm^{-1} = 2954 (w), 1970 (w), 1715 (s), 1438 (m), 1342 (m), 1259 (s), 1194 (m), 1164 (s), 1080 (w), 1025 (m), 984 (w), 851 (m), 777 (w), 724 (w), 700 (w), 458 (vw), 390 (w). The analytical data match those reported in literature[247].

Ethyl buta-2, 3-dienoate (143b)

General procedure 1 (GP 1) was followed by adding Et$_3$N (6.42 mL, 5.07 g, 50.0 mmol, 1.00 equiv.) to a solution of ylide **142b** (17.4 g, 50.0 mmol, 1.00 equiv.) in 150 mL of CH$_2$Cl$_2$, acetyl chloride (4.32 g, 55.0 mmol, 1.00 equiv.) was added dropwise and the reaction was stirred under room temperature for 6 hours. Product **143b** was isolated *via* column chromatography (pentane/Et$_2$O = 3:1) as a colorless liquid, 3.86 g, 34.5 mmol, 69%.

R_f = 0.51 (pentane/Et$_2$O = 4:1). – **^{1}H NMR** (300 MHz, CDCl$_3$) δ/ppm = 5.57 (t, ^4J = 6.5 Hz, 1H, C*H*CO$_2$Et), 5.16 (d, $^2J_{gem}$ = 6.6 Hz, 2H, C*H$_2$*=), 4.14 (q, ^3J = 7.2 Hz, 2H, C*H$_2$*CH$_3$), 1.22 (t, ^3J = 7.1 Hz, 3H, CH$_2$C*H$_3$*). – **^{13}C NMR** (75 MHz, CDCl$_3$) δ/ppm = 215.63 (C$_q$, *C*=CH), 165.63 (C$_q$, *C*O$_2$Et), 88.10 (+, CH, *C*=CH), 79.17 (–, *C*H$_2$=C), 60.88 (–, *C*H$_2$CH$_3$), 13.96 (+, CH$_2$*C*H$_3$). – **IR** (ATR): ṽ/cm^{-1} = 2984 (w), 1969 (w), 1711 (s), 1425 (w), 1368 (w), 1334 (w), 1300 (w), 1253 (m), 1160 (m), 1083 (w), 1035 (m), 915 (w), 851 (m), 777 (w), 731 (m), 648 (w). The analytical data match those reported in literature[248].

tert-Butyl buta-2, 3-dienoate (143c)

General procedure 1 (GP 1) was followed by adding Et$_3$N (2.35 mL, 1.86 g, 13.3 mmol, 1.00 equiv.) to a solution of ylide **142c** (5.00 g, 13.3 mmol, 1.00 equiv.) in 50 mL of CH$_2$Cl$_2$, acetyl chloride (1.15 g, 14.6 mmol, 1.00 equiv.) was added dropwise and the reaction was stirred under room temperature overnight. Product **143c** was isolated *via* column chromatography (pentane/Et$_2$O = 3:1) as a colorless liquid, 1.05 g, 7.5 mmol, 69%.

R_f = 0.55 (*c*-Hex/EtOAc = 3:1). – **^{1}H NMR** (300 MHz, CDCl$_3$) δ/ppm = 5.48 (t, ^4J = 6.5 Hz, 1H, C*H*CO$_2$tBu), 5.10 (d, $^2J_{gem}$ = 6.5 Hz, 2H, C*H*$_2$=), 1.37 (s, 9H, O(C*H*$_3$)$_3$). – **^{13}C NMR** (75 MHz, CDCl$_3$) δ/ppm = 215.29 (C$_q$, *C*=CH), 164.84 (C$_q$, *C*O$_2$tBu), 89.43 (+, CH, C=*C*H), 80.85 (C$_q$, *C*(CH$_3$)$_3$), 78.74 (–, *C*H$_2$=C), 27.92 (+, 3 × *C*H$_3$).The analytical data match those reported in literature[249].

Benzyl buta-2, 3-dienoate (143d)

General procedure 1 (GP 1) was followed by adding Et$_3$N (15.4 mL, 12.2 g, 0.12 mol, 1.00 equiv.) to a solution of ylide **142d** (48.7 g, 0.12 mol, 1.00 equiv.) in 300 mL of CH$_2$Cl$_2$, acetyl chloride (10.2 g, 0.13 mol, 1.10 equiv.) was added dropwise and the reaction was stirred under room temperature overnight. Product **143d** was isolated *via* column chromatography (pentane/Et$_2$O = 3:1) as a colorless liquid, 10.2 g, 58.6 mmol, 49%.

R_f = 0.41 (*c*-Hex/EtOAc = 4:1). – **^{1}H NMR** (300 MHz, CDCl$_3$) δ/ppm = 7.29–7.21 (m, 5H, CH$_2$*Ph*), 5.58 (t, ^4J = 6.5 Hz, 1H, C*H*CO$_2$Bn), 5.10 (d, $^2J_{gem}$ = 6.5 Hz, 2H, C*H*$_2$=), 5.07 (s, 2H, C*H*$_2$Ph). – **^{13}C NMR** (75 MHz, CDCl$_3$) δ/ppm = 216.01 (C$_q$, *C*=CH), 165.56 (C$_q$, *C*O$_2$Bn), 135.86 (C$_q$, *C*^{Ar}), 128.74 (+, CH, 2 × *C*^{Ar}), 128.35 (+, CH, *C*^{Ar}), 127.98 (+, CH, 2 × *C*^{Ar}), 87.96 (+, CH, C=*C*H), 79.49 (–, *C*H$_2$=C), 66.65 (–, *C*H$_2$Ph). – **IR** (ATR): $\tilde{v}$/cm^{-1} = 3033 (vw), 1969 (w), 1711 (s), 1497 (w), 1455 (w), 1422 (w), 1377 (w), 1329 (w), 1244 (m), 1151 (s), 1081 (m), 998 (w), 853 (m), 775 (w), 724 (w), 736 (m), 696 (m), 592 (w), 458 (w). The analytical data match those reported in literature[248].

6.2.6 Synthesis and Characterization of the Unsaturated α-Diazocarbonyl Compound

4-Acetamidobenzenesulfonyl azide (*p*-ABSA)

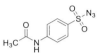

To a solution of *N*-acetylsulfanilyl chloride (23.3 g, 100 mmol, 1.00 equiv.) in 300 mL of acetone (3 mL/mmol *N*-acetylsulfanilyl chloride) and 300 mL of H_2O (3 mL/mmol *N*-acetylsulfanilyl chloride) was added sodium azide (6.50 g, 100 mmol, 1.00 equiv.) carefully at 0 °C. After stirring at 0 °C for 2.5 h, the reaction mixture was concentrated under reduced pressure (30 °C water bath temperature) to half of its volume. The residue was extracted with Et_2O (3 × 50 mL), dried over $MgSO_4$ and the solvent was evaporated under reduced pressure at 30 °C. The product was obtained as light-yellow solid, 21.5 g, 89.6 mmol, 90%.

– **¹H NMR** (300 MHz, $CDCl_3$) δ/ppm = 7.90 (d, *J* = 8.8 Hz, 2H, *CH*^{Ar}), 7.77 (d, *J* = 8.9 Hz, 2H, *CH*^{Ar}), 7.65 (s, 1H, N*H*), 2.25 (s, 3H, CO*CH₃*). – **¹³C NMR** (101 MHz, $CDCl_3$) δ/ppm = 169.22 (C_q, CO*CH₃*), 144.10 (C_q, *C*^{Ar}), 132.09 (C_q, *C*^{Ar}), 128.52 (+, CH, 2 × *C*^{Ar}), 119.58 (+, CH, 2 × *C*^{Ar}), 23.98 (+, CH₃, CO*CH₃*). The analytical data match those reported in the literature.[250]

Dimethyl (*E*)-5-acetylhex-2-enedioate (130aa)

General procedure 2 (GP 2) was followed by adding methyl buta-2,3-dienoate (**143a**, 1.08 g, 11.0 mmol, 1.10 equiv.) dropwise to a suspension of methyl acetoacetate (1.16 g, 10.0 mmol, 1.00 equiv.) and PPh₃ (0.520 g, 2.00 mmol, 20 mol%) in 15 mL of dry benzene. After stirring under reflux for 8 h, the solvent was removed under reduced pressure and the resulting mixture was purified *via* column chromatography (*c*-Hex/EtOAc = 5:1 → 3:1), product **130aa** was obtained as a colorless liquid, 0.910 g, 42.5 mmol, 43%.

R$_f$ = 0.10 (*c*-Hex/EtOAc = 5:1). – **¹H NMR** (400 MHz, $CDCl_3$) δ/ppm = 6.80 (dt, $^3J_{trans}$ = 15.7, ^3J = 7.1 Hz, 1H, CH₂C*H*=CH), 5.84 (dt, $^3J_{trans}$ = 15.6, ^4J = 1.5 Hz, 1H, CH₂CH=C*H*), 3.72 (s, 3H, CH=CHCO₂C*H₃*), 3.68 (s, 3H, CO₂C*H₃*), 3.58 (t, ^3J = 7.3 Hz, 1H, C*H*CH₂), 2.70 (td, ^3J = 7.2, ^3J = 1.5 Hz, 2H, CHC*H₂*), 2.23 (s, 3H, COC*H₃*). – **¹³C NMR** (101 MHz, $CDCl_3$) δ/ppm = 202.06 (C_q, CH₃*C*=O), 169.90 (C_q, *C*O₂CH₃), 167.26, (C_q, *C*O₂CH₃), 145.16 (+, CH, CH₂*C*H=CH), 124.29 (+, CH, CH₂CH=*C*H), 58.76 (+, CH, *C*HCH₂), 53.59 (+, CH₃, CO₂*C*H₃), 52.42 (+, CH₃, CO₂*C*H₃), 31.09 (–, *C*H₂CH),

30.18 (+, CH$_3$, COCH$_3$). – **IR** (ATR): $\tilde{v}$/cm^{-1} = 2955 (w), 1714 (s), 1658 (m), 1435 (m), 1358 (w), 1270 (m), 1198 (m), 1151 (s), 1096 (w), 1034 (m), 981 (m), 862 (w), 712 (w), 603 (vw). – **MS** (EI, 70 eV), m/z (%): 214 (16) [M]$^+$, 171 (26) [M – C$_2$H$_3$O]$^+$, 140 (100) [M – C$_2$H$_3$O – CH$_3$O]$^+$, 109 (27) [C$_6$H$_5$O$_2$]$^+$, 81 (29) [C$_5$H$_5$O]$^+$, 59 (14) [C$_2$H$_3$O$_2$].– **HRMS** (EI, C$_{10}$H$_{14}$O$_5$) calc. 214.0841, found 214.0843.

Dimethyl (*E*)-5-diazohex-2-enedioate (103a)

General procedure 3 (GP 3) was followed by adding KF (0.61 g, 10.50 mmol, 3.00 equiv.) to a solution of dimethyl (*E*)-5-acetylhex-2-enedioate (**130aa**, 0.750 g, 3.50 mmol, 1.00 equiv.) in 15 mL of CH$_3$CN, *p*-ABSA (1.68 g, 7.00 mmol, 2.00 equiv.) in 15 mL of CH$_3$CN was added subsequently, the mixture was stirred at room temperature overnight. Product **103a** was isolated *via* column chromatography (*c*-Hex/EtOAc = 4:1) as a yellow liquid, 0.390 g, 1.97 mmol, 56%.

R_f = 0.34 (*c*-Hex/EtOAc = 3:1). – **^{1}H NMR** (500 MHz, CDCl$_3$) δ/ppm = 6.88 (dt, $^3J_{trans}$ = 15.5, ^3J = 6.4 Hz, 1H, CH$_2$C*H*=CH), 5.90 (dt, $^3J_{trans}$ = 15.7, ^4J = 1.4 Hz, 1H, CH$_2$CH=C*H*), 3.75 (s, 3H, CH=CHCO$_2$C*H$_3$*), 3.71 (s, 3H, CO$_2$C*H$_3$*), 3.17 (dd, ^3J = 6.4, ^4J = 1.7 Hz, 1H, C*H*CH$_2$). –**^{13}C NMR** (126 MHz, CDCl$_3$) δ/ppm = 169.79 (C$_q$, *C*O$_2$CH$_3$), 166.30, (C$_q$, *C*O$_2$CH$_3$), 141.98 (+, CH, CH$_2$*C*H=CH), 123.04 (+, CH, CH$_2$CH=*C*H), 52.90 (+, CH$_3$, CO$_2$*C*H$_3$), 51.66 (+, CH$_3$, CO$_2$*C*H$_3$), 33.85 (–, CH$_2$, *C*H$_2$CH=). – **IR** (ATR): $\tilde{v}$/cm^{-1} = 2954 (w), 2083 (m), 1720 (s), 1658 (m), 1688 (s), 1660 (m), 1435 (m), 1327 (m), 1271 (m), 1192 (m), 1117 (m), 1035 (m), 983 (m), 917 (w), 807 (w), 744 (w), 552 (vw), 453 (vw), 403 (vw). – **MS** (EI, 70 eV), m/z (%): 198 (2) [M]$^+$, 170 (12) [M – N$_2$]$^+$, 111 (100) [M – N$_2$ – C$_2$H$_3$O$_2$]$^+$, 59 (31) [C$_2$H$_3$O$_2$]$^+$. – **HRMS** (EI, C$_8$H$_{10}$O$_4$N$_2$) calc. 198.0641, found 198.0642.

1-Ethyl 6-methyl (*E*)-5-acetylhex-2-enedioate (130ba)

General Procedure 2 (GP 2) was followed by adding ethyl buta-2, 3-dienoate (**143b**, 2.24 g, 20.0 mmol, 1.10 equiv.) dropwise to a solution of methyl acetoacetate (2.32 g, 20.0 mmol, 1.00 equiv.) and PPh$_3$ (1.05 g, 4.00 mmol, 20 mol%) in 40.0 mL of dry benzene. After stirring under reflux for 8 h, the solvent was removed under reduced pressure and the resulting mixture was purified *via* column chromatography (*c*-Hex/EtOAc = 4:1). The product **130ba** was obtained as a colorless liquid, 3.12 g, 13.7 mmol, 68%.

$R_f = 0.33$ (*c*-Hex/EtOAc = 3:1). – **¹H NMR**(400 MHz, CDCl₃) δ/ppm = 6.66 (dt, 1H,³J_{trans} = 15.6, ³J = 7.1 Hz, CH₂C*H*=CH), 5.69 (dt, ³J_{trans} = 15.6, ⁴J = 1.5 Hz, 1H, CH₂CH=C*H*), 3.99 (q, ³J = 7.1 Hz, 2H, OC*H₂*CH₃), 3.58 (s, 3H, OC*H₃*), 3.50 (t, ³J = 7.2 Hz, 1H, C*H*CH₂), 2.74–2.32 (m, 2H, CHC*H₂*), 2.10 (s, 3H, C*H₃*), 1.10 (t, ³J = 7.1 Hz, 3H, OCH₂C*H₃*). – **¹³C NMR** (101 MHz, CDCl₃) δ/ppm = 202.01 (C_q, CH₃*C*O), 169.81 (C_q, *C*O₂CH₃), 166.61 (C_q, *C*O₂Et), 144.86 (+, CH, CH₂*C*H=CH), 124.46 (+, CH, CH₂CH=*C*H), 61.03 (–, O*C*H₂CH₃), 58.53 (+, CH, *C*HCH₂), 53.35 (+, CH₃, CO₂*C*H₃), 30.90 (–, *C*H₂CH), 29.97 (+, CH₃, CO*C*H₃), 14.91 (+, CH₃, OCH₂*C*H₃). – **IR** (ATR): ṽ/cm⁻¹ = 2957 (w), 1712 (s), 1655 (m), 1436 (w), 1366 (m), 1265 (m), 1150 (s), 1095 (m), 1037 (m), 980 (m), 862 (w), 712 (w), 602 (vw), 545 (vw), 460 (vw).– **MS** (EI, 70 eV), m/z (%): 228 (27) [M]⁺, 185 (33) [M – C₂H₃O] ⁺, 183 (25) [M – C₂H₅O]⁺, 155 (31) [M – C₃H₅O₂]⁺, 140 (63) [C₇H₈O₃]⁺, 108 (100) [C₆H₅O₂] ⁺, 81 (33) [C₅H₅O₃]⁺. – **HRMS** (EI, C₁₁H₁₆O₅) calc. 228.0997, found 228.0998.

1-Ethyl 6-methyl (*E*)-5-diazohex-2-enedioate (103b)

General procedure 3 (GP 3) was followed by adding KF (87.0 mg, 1.5 mmol, 3.00 equiv.) to a solution of 1-ethyl 6-methyl (*E*)-5-acetylhex-2-enedioate (**130b**, 114 mg, 0.500 mmol, 1.00 equiv.) in 5 mL of CH₃CN, *p*-ABSA (240 mg, 1.00 mmol 2.00 equiv.) in 5 mL of CH₃CN was added subsequently and the reaction mixture was stirred under room temperature overnight. The product **103b** was isolated *via* column chromatography (*c*-Hex/EtOAc = 6:1) as a yellow liquid, 59.4 mg, 0.280 mmol, 56%.

$R_f = 0.45$ (*c*-Hex/EtOAc = 4:1). – **¹H NMR** (500 MHz, CDCl₃) δ/ppm = 6.90 (dt, ³J_{trans} = 15.6, ³J = 6.4 Hz, 1H, CH₂C*H*=CH), 5.91 (dt, ³J_{trans} = 15.6, ⁴J = 1.6 Hz, 1H, CH₂CH=C*H*), 4.19 (q, ³J = 7.1 Hz, 2H, C*H₂*CH₃), 3.77 (s, 3H, OC*H₃*), 3.18 (dd, ³J = 6.4, ⁴J = 1.6 Hz, 2H, C*H₂*CH=), 1.28 (t, ³J = 7.1 Hz, 3H, CH₂C*H₃*). – **¹³C NMR** (126 MHz, CDCl₃) δ/ppm = 165.89 (C_q, *C*O₂Et), 142.20 (+, CH, CH₂*C*H=CH), 123.50 (+, CH, CH₂CH=*C*H), 60.55 (–, CH₂, O*C*H₂CH₃), 52.19 (+, CH₃, O*C*H₃), 26.11 (–, CH₂, *C*H₂CH=), 14.23 (+, CH₃, OCH₂*C*H₃). – **IR** (ATR): ṽ/cm⁻¹ = 2956 (w), 2083 (m), 1691 (s), 1656 (m), 1437 (m), 1324 (m), 1268 (m), 1162 (s), 1116 (m), 1036 (m), 983 (m), 902 (w), 861 (w), 807 (w), 744 (w), 550 (vw). – **MS** (EI, 70 eV), m/z (%): 212 [M]⁺, 184 (19) [M – N₂]⁺, 111 (100) [M – N₂ – C₃H₅O₂]⁺, 97 (45) [C₂H₃O₂]⁺. – **HRMS** (EI, C₉H₁₂O₄N₂) calc. 212.0797, found 212.0799.

1-Benzyl 6-methyl (*E*)-5-acetylhex-2-enedioate (103ac)

General preoceudre 2 (GP 2) was followed by adding benzyl buta-2,3-dienoate (**143d**, 4.09 g, 23.5 mmol, 1.00 equiv.) dropwise to a solution of methyl acetoacetate (2.73 g, 23.5 mmol, 1.00 equiv.) and PPh$_3$ (1.23 g, 4.70 mmol, 20 mol%) in dry benzene (50.0 mL). After stirring under reflux for 8 h, the solvent was removed under reduced pressure and purified *via* column chromatography (*c*-Hex/EtOAc = 5:1). The product **103ac** was obtained as a colorless liquid, 3.86 g, 13.3 mmol, 57%.

R_f = 0.29 (*c*-Hex/EtOAc = 5:1). – ^{1}H NMR (400 MHz, CDCl$_3$) δ/ppm = 7.37–7.33 (m, 5H, CH$_2$*Ph*), 6.88 (dt, $^3J_{trans}$ = 15.7, ^3J = 7.1 Hz, 1H, CH$_2$C*H*=CH), 5.92 (dt, ^3J = 15.6, ^4J = 1.6 Hz, 1H, CH$_2$CH=C*H*), 5.16 (s, 2H, C*H$_2$*Ph), 3.74 (s, 3H, OC*H$_3$*), 3.60 (t, ^3J = 7.2 Hz, 1H, CH$_2$C*H*), 2.73 (td, ^3J = 7.2, ^4J = 1.5 Hz, 2H, C*H$_2$*CH), 2.25 (s, 3H, C*H$_3$*). – ^{13}C NMR (101 MHz, CDCl$_3$) δ/ppm = 202.05 (C$_q$, CH$_3$*C*=O), 169.91 (C$_q$, *C*O$_2$CH$_3$), 166.67 (C$_q$, *C*O$_2$Bn), 145.58 (+, CH, CH$_2$*C*H=CH), 136.78 (C$_q$, *C*^{Ar}), 129.46 (+, CH, 2 × *C*^{Ar}), 129.13 (+, CH, *C*^{Ar}), 129.14 (+, CH, 2 × *C*^{Ar}), 124.41 (+, CH, CH$_2$CH=*C*H), 67.16 (–, CH$_2$, *C*H$_2$Ph), 58.77 (+, CH, *C*HCH$_2$), 53.64 (+, CH$_3$, CO$_2$*C*H$_3$), 31.14 (–, CH$_2$, *C*H$_2$CH), 30.23 (+, CH$_3$, CO*C*H$_3$). – IR (ATR): ṽ/cm^{-1} = 2954 (vw), 1712 (s), 1655 (m), 1498 (vw), 1436 (w), 1358 (w), 1252 (m), 1149 (s), 980 (m), 857 (w), 738 (m), 697 (m), 602 (vw), 508 (w). – MS (EI, 70 eV), m/z (%): 290 (2) [M]$^+$, 199 (28) [C$_9$H$_{11}$O$_5$]$^+$, 225 (100) [C$_{12}$H$_{17}$O$_4$]$^+$, 213 (12) [C$_{11}$H$_{17}$O$_4$]$^+$, 200 (96) [C$_9$H$_{11}$O$_5$]$^+$, 91 (100) [C$_7$H$_7$]$^+$. – HRMS (EI, C$_{16}$H$_{18}$O$_5$) calc 290.1154, found 290.1154.

1-Benzyl 6-methyl (*E*)-5-diazohex-2-enedioate (103e)

General procedure 3 (GP 3) was followed by adding KF (1.14 g, 19.7 mmol, 3.00 equiv.) to a solution of 1-benzyl 6-methyl (*E*)-5-acetylhex-2-enedioate (**103ac**, 1.90 g, 6.50 mmol, 1.00 equiv.) in 20 mL of CH$_3$CN, *p*-ABSA (3.15 g, 13.1 mmol, 2.00 equiv.) in 10 mL of CH$_3$CN was added subsequently and the reaction mixutre was stirred under room temperature overnight. The product was isolated *via* column chromatography (*c*-Hex/acetone = 6:1) as a yellow liquid, 1.08 g, 0.280 mmol, 56%.

R_f = 0.35 (*c*-Hex/acetone = 5:1). – ^{1}H NMR (400 MHz, CDCl$_3$) δ/ppm = 7.37–7.26 (m, 5H, CH$_2$*Ph*), 6.92 (dt, $^3J_{trans}$ = 15.6, ^3J = 6.4 Hz, 1H, CH$_2$C*H*=CH), 5.93 (dt, $^3J_{trans}$ = 15.6, ^4J = 1.6 Hz, 1H, CH$_2$CH=C*H*), 5.22 (s, 2H, C*H$_2$*Ph), 3.74 (s, 3H, OC*H$_3$*), 3.21 (dd, ^3J = 6.5, ^4J = 1.6 Hz, 2H, C*H$_2$*CH=). – ^{13}C NMR (101 MHz, CDCl$_3$) δ/ppm = 167.19 (C$_q$, *C*O$_2$Me), 143.36 (+, CH, CH$_2$*C*H=CH), 136.72 (C$_q$, *C*^{Ar}), 129.50 (+, CH, 2 × *C*^{Ar}), 129.23 (+, CH, *C*^{Ar}), 129.03 (+, CH, 2 × *C*^{Ar}), 124.07 (+, CH, CH$_2$CH=*C*H), 67.63 (–, CH$_2$, *C*H$_2$Ph), 52.58

(+, CH$_3$, OC*H$_3$*), 27.05 (–, CH$_2$, C*H$_2$*CH=). – **IR** (ATR): ṽ/cm^{-1} = 2953 (w), 2082 (m), 1688 (s), 1656 (m), 1497 (w), 1436 (m), 1326 (m), 1266 (m), 1192 (m), 1157 (s), 1117 (m), 983 (m), 906 (w), 741 (m), 698 (m), 591 (w), 522 (vw). – **MS** (EI, 70 eV), m/z (%): 274 [M]$^+$, 155 (22) [M – N$_2$] $^+$, 111 (100) [M – N$_2$ – C$_7$H$_7$]$^+$, 91 (100) [C$_7$H$_7$]$^+$. – **HRMS** (EI, C$_{14}$H$_{14}$O$_4$N$_2$) calc. 274.0954, found 274.0954.

6-Benzyl 1-methyl (*E*)-5-acetylhex-2-enedioate (130da)

General procedure 2 (GP 2) was followed by adding methyl buta-2,3-dienoate (**143a**, 2.94 g, 30.0 mmol, 1.00 equiv.) dropwise to a solution of benzyl acetoacetate (6.14 g, 30.0 mmol, 1.00 equiv.) and PPh$_3$ (1.57 g, 6.00 mmol, 20 mol%) in 15 mL of dry benzene. After stirring under reflux for 8 h, the solvent was removed under reduced pressure and the resulting mixture was purified *via* column chromatography (*c*-Hex/EtOAc = 5:1). The product was obtained as a colorless liquid, 3.33 g, 11.5 mmol, 38%.

R$_f$ = 0.29 (*c*-Hex/EtOAc = 5:1). – **^{1}H NMR** (300 MHz, CDCl$_3$) δ/ppm = 7.35 (s, 5H, CH$_2$*Ph*), 6.83 (dt, $^3J_{trans}$ = 15.6, ^3J = 7.1 Hz, 1H, CH$_2$C*H*=CH), 5.85 (dt, ^3J = 15.6, ^4J = 1.5 Hz, 1H, CH$_2$CH=C*H*), 5.18 (s, 2H, C*H$_2$*Ph), 3.70 (s, 3H, OC*H$_3$*), 3.63 (t, ^3J = 7.3 Hz, 1H, C*H*CH$_2$), 2.73 (tt, ^3J = 7.2, ^4J = 1.6 Hz, 2H, CHC*H$_2$*), 2.20 (s, 3H, C*H$_3$*). – **^{13}C NMR** (75 MHz, CDCl$_3$) δ/ppm = 201.00 (C$_q$, CH$_3$*C*=O), 168.40 (C$_q$, *C*O$_2$Bn), 166.33 (C$_q$, *C*O$_2$CH$_3$), 144.20 (+, CH, CH$_2$*C*H=CH), 134.96 (C$_q$, *C^{Ar}*), 128.66 (+, CH, 2 × *C^{Ar}*), 128.60 (+, CH, *C^{Ar}*), 128.46 (+, CH, 2 × *C^{Ar}*), 123.46 (+, CH, CH$_2$CH=*C*H), 67.51 (–, CH$_2$, *C*H$_2$Ph), 58.04 (+, CH, *C*HCH$_2$), 51.52 (+, CH$_3$, CO$_2$*C*H$_3$), 30.19 (–, CH$_2$, *C*H$_2$CH), 29.26 (+, CH$_3$, COC*H$_3$*). – **IR** (ATR): ṽ/cm^{-1} = 2951 (vw), 1713 (s), 1658 (m), 1497 (vw), 1435 (w), 1357 (w), 1269 (m), 1214 (m), 1146 (s), 1095 (m), 1027 (w), 973 (m), 909 (m), 848 (m), 739 (m), 697 (m), 596 (vw), 495 (w), 454 (vw), 399 (vw). – **MS** (EI, 70 eV), m/z (%): 290 (3) [M]$^+$, 199 (13) [C$_9$H$_{11}$O$_5$]$^+$, 91 (100) [C$_7$H$_7$]$^+$. – **HRMS** (EI, C$_{16}$H$_{18}$O$_5$) calc. 290.1149, found 290.1148.

1-Benzyl 6-methyl (*E*)-5-diazohex-2-enedioate (103f)

General procedure 3 (GP 3) was followed by adding KF (5.23 g, 90.0 mmol, 3.00 equiv.) to a soluiton of 1-benzyl 6-methyl (*E*)-5-acetylhex-2-enedioate (**130da**, 8.58 g, 30.0 mmol, 1.00 equiv.) in 30 mL of CH$_3$CN, *p*-ABSA (14.4 g, 60.0 mmol 2.00 equiv.) in 20 mL of CH$_3$CN was added subsequently and the reaction mixture was stirred under room temperature overnight. The product **103f** was isolated *via* column chromatography (*c*-Hex/acetone = 5:1) as a yellow liquid, 3.07 g, 11.2 mmol, 37%.

R_f = 0.20 (*c*-Hex/acetone = 3:1). – **¹H NMR** (400 MHz, CDCl₃) δ/ppm = 7.44–7.29 (m, 5H, CH₂*Ph*), 6.96 (dt, $^3J_{trans}$ = 15.5, ^3J = 6.3 Hz, 1H, CH₂C*H*=CH), 5.97 (dt, $^3J_{trans}$ = 15.6, ^4J = 1.6 Hz, 1H, CH₂CH=C*H*), 5.18 (s, 2H, C*H₂*Ph), 3.77 (s, 3H, OC*H₃*), 3.20 (dd, ^3J = 6.3, ^4J = 1.6 Hz, 2H, C*H₂*CH). – **¹³C NMR** (101 MHz, CDCl₃) δ/ppm = 166.56 (C_q, *C*O₂Bn), 143.83 (+, CH, CH₂*C*H=CH), 136.68 (C_q, *C*^Ar), 129.49 (+, CH, 2 × *C*^Ar), 129.38 (+, CH, *C*^Ar), 129.22 (+, CH, 2 × *C*^Ar), 124.02 (+, CH, CH₂CH=*C*H), 67.31 (–, CH₂, *C*H₂Ph), 53.10 (+, CH₃, CO₂*C*H₃), 27.07 (–, CH₂, *C*H₂CH=).

1-*tert*-Butyl 6-methyl (*E*)-5-acetylhex-2-enedioate (130ab)

General procedure 2 (GP 2) was followed by adding *tert*-butyl buta-2,3-dienoate (**143c**, 0.670 g, 4.80 mmol, 1.00 equiv.) dropwise to a solution of methyl acetoacetate (0.560 g, 4.80 mmol, 1.00 equiv.) and PPh₃ (0.250 g, 0.96 mmol, 20 mol%) in 5.0 mL of dry benzene. After stirring under reflux for 8 h, the solvent was removed under reduced pressure and the resulting mixture was purified *via* column chromatography (*c*-Hex/EtOAc = 5:1). The product was obtained as a colorless liquid, 0.510 g, 2.00 mmol, 42%.

R_f = 0.37 (*c*-Hex/EtOAc = 5:1). – **¹H NMR** (400 MHz, CDCl₃) δ/ppm = 6.65 (dt, $^3J_{trans}$ = 15.6, ^3J = 7.1 Hz, 1H, CH₂C*H*=CH), 5.72 (dt, $^3J_{trans}$ = 15.5, ^4J = 1.5 Hz, 1H, CH₂CH=C*H*), 3.69 (s, 3H, OC*H₃*), 3.53 (t, ^3J = 7.3 Hz, 1H, C*H*CH₂), 2.64 (td, ^3J = 7.3, ^4J = 1.5 Hz, 2H, CHC*H₂*), 2.20 (s, 3H, C*H₃*), 1.40 (s, 9H, O(C*H₃*)₃). – **¹³C NMR** (101 MHz, CDCl₃) δ/ppm = 202.25 (C_q, CH₃*C*=O), 170.00 (C_q, *C*O₂CH₃), 166.24 (C_q, *C*O₂tBu), 143.50 (+, CH, CH₂*C*H=CH), 126.45 (+, CH, CH₂CH=*C*H), 81.54 (C_q, *C*(CH₃)₃), 58.89 (+, CH, *C*HCH₂), 53.59 (+, CH₃, CO₂*C*H₃), 30.98 (–, CH₂, *C*H₂CH), 30.23 (+, CH₃, CO*C*H₃), 28.98 (+, CH₃, O(*C*H₃)₃). – **IR** (ATR): $\tilde{v}$/cm⁻¹ = 2978 (w), 1743 (m), 1709 (s), 1642 (w), 1436 (w), 1413 (w), 1366 (m), 1219 (m), 1144 (s), 1050 (w), 987 (w), 854 (w), 818 (m), 749 (w), 543 (vw), 472 (vw). – **MS** (EI, 70 eV). m/z (%): 256 (49) [M]⁺, 241 (20) [C₁₂H₁₇O₅]⁺, 225 (100) [C₁₂H₁₇O₄]⁺, 213 (12) [C₁₁H₁₇O₄]⁺, 200 (96) [C₉H₁₁O₅]⁺, 183 (21) [C₉H₁₁O₄]⁺, 141 (35) [C₈H₁₃O₂]⁺ 57 (100) [C₄H₉]⁺. – **HRMS** (EI, C₁₃H₂₀O₅) calc. 256.1305, found 256.1303.

1-*tert*-Butyl 6-methyl (*E*)-5-diazohex-2-enedioate (103c)

General procedure 3 (GP 3) was followed by adding KF (0.35 g, 6.00 mmol, 3.00 equiv.) to a solution of 1-*tert*-butyl 6-methyl (*E*)-5-acetylhex-2-enedioate (**130ab**, 0.510 g, 2.00 mmol, 1.00 equiv.) in 10 mL of CH_3CN, *p*-ABSA (0.96 g, 4.00 mmol 2.00 equiv.) in 5 mL of CH_3CN was added subsequently and the reaction mixture was stirred at room temperature overnight. The product was isolated *via* column chromatography (*c*-Hex/acetone = 5:1) as a yellow liquid, 0.290 g, 1.20 mmol, 60%.

R_f = 0.36 (*c*-Hex/EtOAc = 3:1). – **^{1}H NMR** (400 MHz, CDCl$_3$) δ/ppm = 6.79 (dt, $^3J_{trans}$ = 15.5, ^4J = 6.3 Hz, 1H, CH$_2$C*H*=CH), 5.85 (dt, $^3J_{trans}$ = 15.5, ^4J = 1.6 Hz, 1H, CH=C*H*CO$_2t$Bu), 3.77 (s, 3H, OC*H$_3$*), 3.16 (dd, ^3J = 6.3, ^4J = 1.6 Hz, 2H, C*H$_2$*CH), 1.47 (s, 9H, O(C*H$_3$*)$_3$).– **^{13}C NMR** (101 MHz, CDCl$_3$) δ/ppm = 166.09 (C$_q$, *C*O$_2t$Bu), 141.80 (+, CH, CH$_2$*C*H=CH), 126.12 (+, CH, *C*HCO$_2t$Bu), 81.56 (C$_q$, *C*(CH$_3$)$_3$), 53.05 (+, CH$_3$, CO$_2$*C*H$_3$), 28.99 (+, CH$_3$, O(*C*H$_3$)$_3$), 26.84 (–, CH$_2$, *C*H$_2$CH=).

6-*tert*-Butyl 1-methyl (*E*)-5-acetylhex-2-enedioate (130ca)

General procedure 2 (GP 2) was followed by adding methyl buta-2,3-dienoate (**143a**, 0.980 g, 10.0 mmol, 1.00 equiv.) dropwise to a solution of *tert*-butyl acetoacetate (1.58 g, 10.0 mmol, 1.00 equiv.) and PPh$_3$ (0.52 g, 2.0 mmol, 20 mol%) in 10 mL of dry benzene. After stirring under reflux for 8 h, the solvent was removed under reduced pressure and the resulting mixture was purified *via* column chromatography (*c*-Hex/EtOAc = 5:1). The product was obtained as a colorless liquid, 0.840 g, 3.30 mmol, 33%.

R_f = 0.33 (*c*-Hex/EtOAc = 4:1). – **^{1}H NMR** (400 MHz, CDCl$_3$) δ/ppm = 6.79 (dt, $^3J_{trans}$ = 15.7, ^3J = 7.1 Hz, 1H, CH$_2$C*H*=CH), 5.80 (dt, ^3J = 15.6, ^4J = 1.5 Hz, 1H, CH=C*H*CO$_2$CH$_3$), 3.65 (s, 3H, OC*H$_3$*), 3.42 (t, ^3J = 7.3 Hz, 1H, C*H*CH$_2$), 2.70–2.47 (m, 2H, CHC*H$_2$*), 2.18 (s, 3H, C*H$_3$*), 1.39 (s, 9H, O(C*H$_3$*)$_3$). – **^{13}C NMR** (101 MHz, CDCl$_3$) δ/ppm = 202.51 (C$_q$, CH$_3$*C*O), 168.59 (C$_q$, *C*O$_2t$Bu), 167.34 (C$_q$, *C*O$_2$CH$_3$), 145.64 (+, CH, CH$_2$*C*H=CH), 124.02 (+, CH, =*C*HCO$_2$CH$_3$), 83.51 (C$_q$, *C*(CH$_3$)$_3$), 60.00 (+, CH, *C*HCH$_2$), 52.39 (+, CH$_3$, CO$_2$*C*H$_3$), 31.0 9 (–, CH$_2$, *C*H$_2$CH), 30.01 (+, CH$_3$, CO*C*H$_3$), 28.74 (+, CH$_3$, O(*C*H$_3$)$_3$). – **IR** (ATR): $\tilde{\nu}$/cm^{-1} = 2980 (w), 1713 (s), 1659 (w), 1435 (w), 1394 (w), 1368 (m), 1269 (m), 1140 (s), 1037 (w), 970 (w), 841 (w), 718 (w), 599 (vw), 425 (vw). – **MS** (EI, 70 eV), m/z (%): 256 (1) [M]$^+$, 200 (55) [M – C$_4$H$_9$]$^+$, 183 (16) [M – C$_4$H$_9$O]$^+$,

157 (14) $[C_8H_{13}O_3]^+$, 57 (100) $[C_4H_9]^+$. – **HRMS** (EI, $C_{13}H_{20}O_5$) calc. 256.1311, found 256.1309.

6-*tert*-Butyl 1-methyl (*E*)-5-diazohex-2-enedioate (103d)

General procedure 3 (GP 3) was followed by adding KF (0.30 g, 5.10 mmol 3.00 equiv.) to a solution of 6-*tert*-butyl 1-methyl (*E*)-5-acetylhex-2-enedioate (**130ca**, 0.65 g, 2.50 mmol, 1.00 equiv.) in 10 mL of CH_3CN, *p*-ABSA (1.83 g, 7.62 mmol 2.00 equiv.) in 10 mL of CH_3CN was added subsequently and the reaction mixture was stirred at room temperature overnight. The product was isolated *via* column chromatography (*c*-Hex/acetone = 5:1) as a yellow liquid, 0.230 g, 1.00 mmol, 40%.

R_f = 0.50 (*c*-Hex/EtOAc = 3:1). – **^{1}H NMR** (400 MHz, CDCl$_3$) δ/ppm = 6.90 (dt, $^3J_{trans}$ = 15.6, ^3J = 6.5 Hz, 1H, CH$_2$C*H*=CH), 5.90 (dt, $^3J_{trans}$ = 15.6, ^3J = 1.6 Hz, 1H, CH=C*H*CO$_2$CH$_3$), 3.72 (s, 3H, OC*H$_3$*), 3.13 (dd, ^3J = 6.5, ^4J = 1.6 Hz, 2H, CHC*H$_2$*), 1.46 (s, 9H, O(C*H$_3$*)$_3$). – **^{13}C NMR** (101 MHz, CDCl$_3$) δ/ppm = 167.27 (C$_q$, *C*O$_2$Me), 143.81 (+, CH, CH$_2$*C*H=CH), 123.75 (+, CH, *C*HCO$_2$CH$_3$), 82.62 (C$_q$, *C*(CH$_3$)$_3$), 52.52 (+, CH$_3$, CO$_2$*C*H$_3$), 29.22 (+, CH$_3$, O(*C*H$_3$)$_3$), 27.02 (−, CH$_2$, *C*H$_2$CH=).

6.2.7 Synthesis and Characterization of N–H Insertion Products from the Unsaturated *α*-Diazocarbonyl Substrates[VII]

Dimethyl (*E*)-5-([1, 1'-biphenyl]-2-ylamino)hex-2-enedioate (177ab)

General procedure 4a (GP 4a) was followed by stirring Py-box (**172**, 0.21 mg, 6.00 µmol, 6 mol%), NaBArF (5.20 mg, 6.00 µmol, 6 mol%) and Cu(MeCN)$_4$PF$_6$ (1.60 mg, 5.00 µmol, 5 mol%) in abs. CH$_2$Cl$_2$ (1 mL) in an oven dried vial at 40 °C for 2 h. Dimethyl (*E*)-5-diazohex-2-enedioate (**103a**, 19.9 mg, 0.100 mmol, 1.00 equiv.) and 2-aminobiphenyl (20.3 mg, 0.120 mmol, 1.20 equiv.) in 1 mL of CH$_2$Cl$_2$ was added subsequently, the mixture was stirred at room temperature for 2 h. The resulting mixture was purified directly *via* column

---

[VII] The signal of N*H* can not be seen on ^{1}H NMR spectrum.

chromatography (*c*-Hex/EtOAc = 5:1) to give the title product **177ab** as yellow powder, 15.7 mg, 46.0 μmol, 46%.

R_f = 0.46 (*c*-Hex/EtOAc = 4:1). – **^{1}H NMR** (500 MHz, MeOD) δ/ppm = 7.44–7.41 (m, 2H, 2 × CH^{Ar}), 7.39–7.33 (m, 3H, 3 × CH^{Ar}), 7.17 (t, ^3J = 7.7 Hz, 1H, CH^{Ar}), 7.05 (d, ^3J = 7.5 Hz, 1H, CH^{Ar}), 6.82–6.76 (m, 2H, CH^{Ar} + CH_2CH=CH), 6.67 (d, ^3J = 8.2 Hz, 1H, CH^{Ar}), 5.79 (dt, $^3J_{trans}$ = 15.4, ^4J = 1.6 Hz, 1H, CH=CHCO$_2$Me), 4.34 (t, ^3J = 6.0 Hz, 1H, CHN), 3.71 (s, 3H, OCH_3), 3.69 (s, 3H, OCH_3), 2.71–2.59 (m, 2H, CH_2). – **^{13}C NMR** (126 MHz, MeOD) δ/ppm = 174.79 (C$_q$, CO_2CH$_3$), 167.83 (C$_q$, =CHCO$_2$CH$_3$), 144.83 (+, CH, CH$_2$$C$H=CH), 144.55 (C$_q$, C^{Ar}), 140.57 (C$_q$, C^{Ar}), 131.36 (+, CH, C^{Ar}), 130.36 (+, CH, 2 × C^{Ar}), 130.12 (C$_q$, C^{Ar}), 129.93 (+, CH, 2 × C^{Ar}), 129.65 (+, CH, C^{Ar}), 128.40 (+, CH, CH=CHCO$_2$Me), 125.26 (+, CH, C^{Ar}), 119.42 (+, CH, C^{Ar}), 112.69 (+, CH, C^{Ar}), 56.49 (+, CH, CHN), 52.73 (+, CH$_3$, OCH$_3$), 52.05 (+, CH$_3$, 2C, OCH$_3$), 35.81 (–, CH$_2$, CHCH$_2$). – **IR** (ATR): $\tilde{\nu}$/cm^{-1} = 3400 (vw), 2950 (vw), 1721 (m), 1656 (w), 1602 (w), 1581 (w), 1510 (w), 1490 (w), 1434 (w), 1314 (w), 1269 (w), 1198 (w), 1168 (m), 1098 (w), 1034 (w), 1008 (w), 981 (w), 848 (vw), 770 (w), 748 (w), 703 (w), 615 (vw), 469 (vw), 395 (vw). – **MS** (EI, 70 eV), m/z (%): 339 (13) [M]$^+$, 240 (18) [C$_{15}$H$_{14}$NO$_2$]$^+$, 180 (100) [C$_{13}$H$_{11}$N – H]$^+$, – **HRMS** (EI, C$_{20}$H$_{21}$O$_4$N$_1$) calc. 339.1465, found 339.1463.

Dimethyl (*E*)-5-((4-methoxyphenyl)amino)hex-2-enedioate (177ac)

General procedure 4a (GP 4a) was followed by stirring Py-box (**172**, 2.21 mg, 6.00 μmol, 6 mol%), NaBArF (5.20 mg, 6.00 μmol, 6 mol%) and Cu(MeCN)$_4$PF$_6$ (1.60 mg, 5.00 μmol, 5 mol%) in abs. CH$_2$Cl$_2$ (1 mL) in an oven dried vial at 40 °C for 2 h. Dimethyl (*E*)-5-diazohex-2-enedioate (**103a**, 19.8 mg, 0.100 mmol, 1.00 equiv.) in 1 mL of CH$_2$Cl$_2$ and *p*-anisidine (15.0 mg, 0.120 mmol, 1.20 equiv.) were added susequently and the mixture was stirred at room temperature for 2 h. The resulting mixture was purified directly *via* column chromatography (*c*-Hex/EtOAc = 4:1) to give the title product **177ac** as colorless oil, 18.0 mg, 61.0 μmol, 61%.

R_f = 0.19 (*c*-Hex/EtOAc = 4:1). – **^{1}H NMR** (500 MHz, MeOD) δ/m = 6.98 (dt, $^3J_{trans}$ = 15.6, ^3J = 7.3 Hz, 1H, CH$_2$CH=CH), 6.79–6.71 (m, 2H, 2 × CH^{Ar}), 6.66–6.56 (m, 2H, 2 × CH^{Ar}), 5.95 (dt, $^3J_{trans}$ = 15.6, ^4J = 1.5 Hz, 1H, CH=CHCO$_2$Me), 4.17 (t, ^3J = 6.0 Hz, 1H, CHN), 3.72 (s, 3H, OCH_3), 3.71 (s, 3H, OCH_3), 3.68 (s, 3H, OCH_3), 2.85–2.53 (m, 2H, CH_2CH). – **^{13}C NMR** (126 MHz, MeOD) δ/ppm = 175.42 (C$_q$, CO_2CH$_3$), 168.20 (C$_q$, =CHCO$_2$CH$_3$), 154.13 (C$_q$, C^{Ar}), 145.76 (+, CH, CH$_2$$C$H=CH), 142.49 (C$_q$, C^{Ar}), 124.50 (+, CH, CH=CHCO$_2$Me), 116.19 (+, CH, 2 × C^{Ar}), 115.76 (+, CH, 2 × C^{Ar}),

58.13 (+, CH, CHN), 56.08 (+,CH_3, OCH_3) 52.49 (+, CH_3, OCH_3), 52.02 (+, CH_3, OCH_3), 36.14 (–, CH_2, CHCH_2). – **IR** (ATR): $\tilde{v}$/cm^{-1} = 3371 (vw), 2951 (vw), 2834 (vw), 2111 (vw), 1719 (m), 1658 (w), 1512 (m), 1435 (w), 1236 (m), 1202 (w), 1167 (m), 1106 (w), 1034 (w), 854 (w), 980 (w), 820 (w), 756 (vw), 706 (vw), 515 (vw). – **MS** (EI, 70 eV), m/z (%): 293 (34) [M]$^+$, 234 (18) [M – $C_2H_3O_2$]$^+$, 194 (100) [$C_{10}H_{12}NO_3$]$^+$, 134 (98) [C_8H_9NO – H]. – **HRMS** (EI, $C_{15}H_{19}O_5N_1$) calc. 293.1258, found 293.1258.

Dimethyl (*E*)-5-((2-methoxyphenyl)amino)hex-2-enedioate (177ad)

General procedure 4a (GP 4a) was followed by stirring Py-box (**172**, 2.21 mg, 6.00 μmol, 6 mol%), NaBArF (5.20 mg, 6.00 μmol, 6 mol%) and Cu(MeCN)$_4$PF$_6$ (1.6 mg, 5.00 μmol, 5 mol%) in abs. CH_2Cl_2 (1 mL) in an oven dried vial at 40 °C for 2 h. Dimethyl (*E*)-5-diazohex-2-enedioate (**103a**, 19.8 mg, 0.100 mmol, 1.00 equiv.) in 1 mL of CH_2Cl_2 and *o*-anisidine (13.7 μL, 15.0 g, 0.120 mmol, 1.20 equiv.) were added subsequently and the mixture was stirred at room temperature for 2 h. The resulting mixture was purified directly *via* column chromatography (*c*-Hex/EtOAc = 5:1) to give the title product **177ad** as colorless oil, 12.2 mg, 42.0 μmol, 42%.

R_f = 0.28 (*c*-Hex/EtOAc = 4:1). – ^{1}**H NMR** (500 MHz, MeOD) δ/ppm = 6.94 (dt, $^3J_{trans}$ = 15.6, ^3J = 7.4 Hz, 1H, CH_2CH=CH), 6.84 (dd, ^3J = 8.0, ^4J = 1.4 Hz, 1H, CH^{Ar}), 6.79 (td, ^3J = 7.7, ^4J = 1.4 Hz, 1H, CH^{Ar}), 6.69 (td, ^3J = 7.7, ^4J = 1.5 Hz, 1H, CH^{Ar}), 6.59 (dd, ^3J = 7.9, ^4J = 1.5 Hz, 1H, CH^{Ar}), 5.93 (dt, $^3J_{trans}$ = 15.6, ^4J = 1.4 Hz, 1H, CH=CHCO$_2$Me), 4.32 (t, ^3J = 6.3 Hz, 1H, CHN), 3.85 (s, 3H, OCH_3), 3.72 (s, 3H, OCH_3), 3.71 (s, 3H, OCH_3), 2.98–2.55 (m, 2H, CH_2CH). – ^{13}**C NMR** (125 MHz, MeOD) δ/ppm = 174.94 (C$_q$, CO_2CH$_3$), 168.09 (C$_q$, =CHCO_2CH$_3$), 148.80 (C$_q$, C^{Ar}), 145.34 (+, CH, CH_2CH=), 137.54 (C$_q$, C^{Ar}), 124.78 (+, CH, =CHCO$_2$Me), 112.05 (+, CH, 2 × C^{Ar}), 111.17 (+, CH, 2 × C^{Ar}), 56.66 (+, CH, CHN), 56.07 (+, CH$_3$, OCH_3), 52.68 (+, CH$_3$, OCH_3), 52.04 (+, CH$_3$, OCH_3), 35.94 (–, CH$_2$, CH_2CH). – **IR** (ATR):$\tilde{v}$/cm^{-1} = 3387 (vw), 2950 (w), 1719 (m), 1657 (w), 1601 (w), 1512 (m), 1456 (w), 1433 (w), 1320 (w), 1249 (w), 1219 (m), 1168 (m), 1025 (m), 980 (w), 849 (vw), 737 (w), 417 (vw). – **MS** (EI, 70 eV), m/z (%): 293 (24) [M]$^+$, 234 (14) [M – $C_2H_3O_2$]$^+$, 194 (100) [$C_{10}H_{12}NO_3$]+, 134 (64) [C_8H_9NO – H]. – **HRMS** (EI, $C_{15}H_{19}O_5N_1$) calc. 293.1258, found 293.1257.

6-Benzyl 1-methyl (*E*)-5-(phenylamino)hex-2-enedioate (176e)

General procedure 4a (GP 4a) was followed by stirring Py-box (**172**, 2.21 mg, 6.00 μmol, 6 mol%), NaBArF (5.20 mg, 6.00 μmol, 6 mol%) and Cu(MeCN)$_4$PF$_6$ (1.60 mg, 5.00 μmol, 5 mol%) in abs. CH$_2$Cl$_2$ (1 mL) in an oven dried vial at 40 °C for 2 h. 6-Benzyl 1-methyl (*E*)-5-diazohex-2-enedioate (**103f**, 27.4 mg, 0.100 mmol, 1.00 equiv.) in 1 mL of CH$_2$Cl$_2$ and aniline (11.2 mg, 0.120 mmol, 1.20 equiv.) were added subsequently and the mixture was stirred at room temperature for 2 h. The resulting mixture was purified directly *via* column chromatography (*c*-Hex/EtOAc = 8:1) to give the product **176e** as colorless oil, 14.1 mg, 42.0 μmol, 42%.

R_f = 0.13 (*c*-Hex/EtOAc = 8:1). – **^{1}H NMR** (400 MHz, CDCl$_3$) δ/ppm = 7.39–7.26 (m, 5H, CH$_2$*Ph*), 7.18 (dd, ^3J = 8.6, ^3J = 7.3 Hz, 2H, 2 × C*H*^{Ar}), 6.89 (dt, $^3J_{trans}$ = 15.2, ^3J = 7.3 Hz, 1H, CH$_2$C*H*=CH), 6.80–6.72 (m, 1H, C*H*^{Ar}), 6.65–6.57 (m, 2H, 2 × C*H*^{Ar}), 5.87 (dt, $^3J_{trans}$ = 15.6, ^4J = 1.5 Hz, 1H, CH$_2$CH=C*H*), 5.16 (d, J = 3.1 Hz, 2H, C*H$_2$*Ph), 4.29 (t, ^3J = 6.1 Hz, 1H, C*H*N), 3.72 (s, 3H, OC*H$_3$*), 2.84–2.57 (m, 2H, C*H$_2$*CH=CH). – **^{13}C NMR** (101 MHz, CDCl$_3$) δ/ppm = 173.43 (C$_q$, *C*O$_2$Bn), 167.10 (C$_q$, *C*O$_2$Me), 146.91 (C$_q$, *C*^{Ar}), 143.69 (+, CH, CH$_2$CH=), 136.04 (C$_q$, *C*^{Ar}), 130.32 (+, CH, 2 × *C*^{Ar}), 129.53 (+, CH, 2 × *C*^{Ar}), 129.43 (+, CH, *C*^{Ar}), 129.34 (+, CH, 2 × *C*^{Ar}), 125.33 (+, CH, CH$_2$CH=*C*H), 119.73 (+, CH, *C*^{Ar}), 114.58 (+, CH, 2 × *C*^{Ar}), 68.18 (–, CH$_2$, *C*H$_2$Ph), 56.56 (+, CH, *C*HN), 52.49 (+, CH$_3$, O*C*H$_3$), 36.13 (–, CH$_2$, *C*H$_2$CH=CH). – **IR** (ATR): $\tilde{v}$/cm^{-1} = 3371 (vw), 2950 (w), 2113 (vw), 1719 (s), 1658 (w), 1602 (m), 1498 (m), 1435 (m), 1379 (vw), 1314 (m), 1266 (m), 1165 (s), 1082 (w), 1029 (m), 976 (m), 911 (w), 846 (w), 748 (m), 693 (m), 500 (w). – **MS** (EI, 70 eV), m/z (%): 339 (5) [M]$^+$, 240 (21) [C$_{15}$H$_{14}$NO$_2$]$^+$, 204 (10) [C$_{12}$H$_{14}$NO$_2$]$^+$, 91 (100) [C$_7$H$_7$]$^+$. – **HRMS** (EI, C$_{20}$H$_{21}$O$_4$N$_1$) calc. 339.1471, found 339.1471.

6-Benzyl 1-methyl (*E*)-5-((4-methoxyphenyl)amino)hex-2-enedioate (177fc)

General procedure 4a (GP 4a) was followed by using Py-box (**172**, 2.21 mg, 6.00 μmol, 6 mol%), NaBArF (5.20 mg, 6.00 μmol, 6 mol%) and Cu(MeCN)$_4$PF$_6$ (1.60 mg, 5.00 μmol, 5 mol%) in abs. CH$_2$Cl$_2$ (1 mL) in an oven dried vial at 40 °C for 2 h. 6-Benzyl 1-methyl (*E*)-5-diazohex-2-enedioate (**103f**, 27.4 mg, 0.100 mmol, 1.00 equiv.) in 1 mL of CH$_2$Cl$_2$ and *p*-anisidine (15.0 mg, 0.120 mmol, 1.20 equiv.) were added subsequently and the mixture was stirred at room temperature for 2 h. The resulting mixture was purified directly *via* column chromatography (*c*-Hex/EtOAc = 6:1) to give the title product **177fc** as colorless oil, 16.0 mg, 43.0 μmol, 43%.

R_f = 0.13 (*c*-Hex/EtOAc = 8:1). – **^{1}H NMR** (400 MHz, MeOD) δ/ppm = 7.32–7.19 (m, 5H, CH$_2$*Ph*), 6.94 (dt, $^3J_{trans}$ = 15.7, ^3J = 7.3 Hz, 1H, CH$_2$C*H*=), 6.74–6.67 (m, 2H, 2 × C*H*^{Ar}), 6.64–6.55 (m, 2H, 2 × C*H*^{Ar}), 5.88 (dt, $^3J_{trans}$ = 15.6, ^4J = 1.5 Hz, 1H, =C*H*CO$_2$Me), 5.19–4.99 (m, 2H, C*H*$_2$Ph), 4.18 (t, ^3J = 6.8 Hz, 1H, C*H*N), 3.69 (s, 3H, OC*H*$_3$), 3.68 (s, 3H, OC*H*$_3$), 2.68 (qd, ^3J = 7.4, ^4J = 1.5 Hz, 2H, C*H*$_2$CH=). – **^{13}C NMR** (101 MHz, MeOD) δ/ppm = 174.80 (C$_q$, *C*O$_2$Bn), 168.16 (C$_q$, *C*O$_2$CH$_3$), 154.20 (C$_q$, *C*^{Ar}), 145.60 (+, CH, CH$_2$*C*H=), 142.48 (C$_q$, *C*$_{Ar}$), 137.18 (C$_q$, *C*^{Ar}), 129.51 (+, CH, 2 × *C*^{Ar}), 129.39 (+, CH, 2 × *C*^{Ar}), 129.28 (+, CH, *C*^{Ar}), 124.60 (+, CH, =*C*HCO$_2$Me), 116.41 (+, CH, 2 × *C*^{Ar}), 115.81 (+, CH, 2 × *C*^{Ar}), 67.79 (–, CH$_2$, *C*H$_2$Ph), 53.34 (+, CH, *C*HN), 56.14 (+, CH$_3$, O*C*H$_3$), 52.03 (+, CH$_3$, *C*O$_2$CH$_3$), 36.17 (–, CH$_2$, *C*H$_2$).– **IR** (ATR): ṽ/cm^{-1} = 3371 (w), 2947 (w), 2836 (vw), 1717 (m), 1659 (w), 1509 (m), 1431 (w), 1380 (vw), 1363 (w), 1317 (w), 1283 (w), 1254 (m), 1220 (m), 1180 (m), 1166 (m), 1104 (w), 1082 (w), 1037 (m), 986 (m), 925 (w), 909 (w), 822 (m), 808 (w), 757 (w), 731 (m), 694 (m), 595 (w), 556 (w). – **MS** (EI, 70 eV), m/z (%): 369 (27) [M]$^+$, 270 (55) [C$_{16}$H$_{16}$NO$_3$]$^+$, 134 (99) [C$_8$H$_8$NO]$^+$, 91 (100) [C$_6$H$_5$N]$^+$, 59 (13) [C$_2$H$_3$O$_2$]$^+$. – **HRMS** (EI, C$_{21}$H$_{23}$O$_4$N$_1$O$_5$) calc. 369.1574, found 369.1576.

1-Benzyl 6-methyl (*E*)-5-((4-methoxyphenyl)amino)hex-2-enedioate (177ec)

General procedure 4a (GP 4a) was followed by stirring Py-box (**172**, 2.21 mg, 6.00 μmol, 6 mol%), NaBArF (5.20 mg, 6.00 μmol, 6 mol%) and Cu(MeCN)$_4$PF$_6$ (1.60 mg, 5.00 μmol, 5 mol%) in abs. CH$_2$Cl$_2$ (1 mL) in an oven dried vial at 40 °C for 2 h. 1-Benzyl 6-methyl (*E*)-5-diazohex-2-enedioate (**103e**, 27.4 mg, 0.100 mmol, 1.00 equiv.) in 1 mL of CH$_2$Cl$_2$ and *p*-anisidine (15.0 mg, 0.120 mmol, 1.20 equiv.) were added subsequently and the mixture was stirred at room temperature for 2 h. The resulting mixture was purified directly *via* column chromatography (*c*-Hex/EtOAc = 8:1 → 6:1) to give the title product **177ec** as colorless oil, 16.2 mg, 0.044 mmol, 44%.

R_f = 0.22 (*c*-Hex/EtOAc = 4:1). – **^{1}H NMR** (400 MHz, MeOD) δ/ppm = 7.37–7.26 (m, 5H, CH$_2$*Ph*), 6.99 (dt, $^3J_{trans}$ = 15.1, ^3J = 7.8 Hz, 1H, CH$_2$C*H*=), 6.77–6.69 (m, 2H, 2 × C*H*^{Ar}), 6.65–6.58 (m, 2H, 2 × C*H*^{Ar}), 5.97 (dt, $^3J_{trans}$ = 15.1, ^3J = 1.5 Hz, 1H, =C*H*CO$_2$Bn), 5.15 (s, 2H, C*H*$_2$Ph), 4.15 (t, ^3J = 6.2 Hz, 1H, C*H*N), 3.68 (s, 3H, CO$_2$C*H*$_3$), 3.65 (s, 3H, OC*H*$_3$), 2.68 (qd, ^3J = 7.4, ^4J = 1.5 Hz, 2H, C*H*$_2$CH=). – **^{13}C NMR** (101 MHz, MeOD) δ/ppm = 175.43 (C$_q$, *C*O$_2$CH$_3$), 167.47 (C$_q$, *C*O$_2$Bn), 154.16 (C$_q$, *C*^{Ar}), 146.12 (+, CH,

CH$_2$CH=), 142.49 (C$_q$, C^{Ar}), 137.61 (C$_q$, C^{Ar}), 129.56 (+, CH, 2 × C^{Ar}), 129.21 (+, CH, 3 × C^{Ar}), 124.68 (+, CH, =CHCO$_2$Me), 116.26 (+, CH, 2 × C^{Ar}), 115.82 (+, CH, 2 × C^{Ar}), 67.23 (–, CH$_2$, CH_2Ph), 58.16 (+, CH, CHN), 56.13 (+, CH$_3$, OCH$_3$), 52.52 (+, CH$_3$, CO$_2$CH$_3$), 36.26 (–, CH$_2$, CH_2CH=). – **IR** (ATR): ṽ/cm^{-1} = 3368 (vw), 2951 (w), 1715 (m), 1655 (w), 1511 (s), 1468 (w), 1377 (w), 1236 (s), 1209 (m), 1162 (s), 1034 (m), 979 (m), 821 (m), 739 (m), 697 (m), 591 (w), 520 (w). – **MS** (EI, 70 eV), m/z (%): 369 (26) [M]$^+$, 194 (100) [C$_{10}$H$_{12}$NO$_3$]$^+$, 135 (11) [C$_8$H$_7$O$_2$]$^+$, 134 (99) [C$_8$H$_8$NO]$^+$, 122 (11) [C$_7$H$_8$NO]$^+$, 107 (14) [C$_7$H$_7$O]$^+$, 91 (48) [C$_6$H$_5$N]$^+$, 59 (13) [C$_2$H$_3$O$_2$]$^+$. – **HRMS** (EI, C$_{20}$H$_{21}$O$_4$N) calc. 369.1576, found 369.1578.

6-Benzyl 1-methyl (*E*)-5-((2-chlorophenyl)amino)hex-2-enedioate (177ff)

General procedure 4a (GP 4a) was followed by stirring Py-box (**172**, 2.21 mg, 6.00 μmol, 6 mol%), NaBArF (5.20 mg, 6.00 μmol, 6 mol%) and Cu(MeCN)$_4$PF$_6$ (1.60 mg, 5.00 μmol, 5 mol%) in abs. CH$_2$Cl$_2$ (1 mL) in an oven dried vial at 40 °C for 2 h. 6-Benzyl 1-methyl (*E*)-5-diazohex-2-enedioate (**103f**, 27.4 mg, 0.100 mmol, 1.00 equiv.) in 1 mL of CH$_2$Cl$_2$ and 2-chloroaniline (12.6 μL, 0.120 mmol, 1.20 equiv.) were added subsequently and the mixture was stirred at room temperature for 2 h. The resulting mixture was purified directly *via* column chromatography (pentane/Et$_2$O = 5:1) to give the title product **177ff** as colorless oil, 20.3 mg, 54.0 μmol, 54%.

R_f = 0.36 (pentane/Et$_2$O = 5:1). – **^{1}H NMR** (400 MHz, CDCl$_3$) δ/ppm = 7.40–7.32 (m, 5H, CH_2Ph), 7.29–7.26 (m, 1H, CH^1), 7.09 (ddd, ^3J = 8.1, ^3J = 7.4, ^4J = 1.5 Hz, 1H, CH^3), 6.89 (dt, $^3J_{trans}$ = 15.6, ^3J = 7.4 Hz, 1H, CH$_2$CH=CH), 6.69 (td, ^3J = 7.6, ^4J = 1.4 Hz, 1H, CH^4), 6.56 (dd, ^3J = 8.2, ^4J = 1.4 Hz, 1H, CH^2), 5.89 (dt, $^3J_{trans}$ = 15.6, ^4J = 1.5 Hz, 1H, CH$_2$CH=CH), 5.18 (d, J = 4.1 Hz, 2H, CH_2Ph), 4.29 (t, ^3J = 5.5 Hz, 1H, CHN), 3.72 (s, 3H, OCH_3), 3.05–2.29 (m, 2H, CH_2CH=CH). – **^{13}C NMR** (101 MHz, CDCl$_3$) δ/ppm = 172.79 (C$_q$, CO_2Bn), 167.03 (C$_q$, CO_2Me), 143.24 (+, CH, CH$_2$CH=CH), 142.99 (C$_q$, C^{Ar}-NH), 135.97 (C$_q$, C^{Ar}), 130.45 (+, CH, $C_1{}^{Ar}$), 129.54 (+, CH, 2 × C^{Ar}), 129.46 (+, CH, C^{Ar}), 129.34 (+, CH, 2 × C^{Ar}), 128.70 (+, CH, CH$_2$CH=CH), 125.53 (+, CH, $C_3{}^{Ar}$), 120.99 (C$_q$, C^{Ar}-Cl), 119.61 (+, CH, $C_2{}^{Ar}$), 112.73 (+, CH, $C_4{}^{Ar}$), 68.29 (–, CH$_2$, CH_2Ph), 56.36 (+, CH, CHN), 52.52 (+, CH$_3$, OCH$_3$), 36.06 (–, CH_2CH=). – **IR** (ATR): ṽ/cm^{-1} = 3401 (vw), 2950 (w), 2107 (vw), 1718 (s), 1658 (w), 1597 (m), 1508 (m), 1456 (w), 1435 (m), 1378 (w), 1319 (m), 1270 (m), 1240 (m), 1166 (m), 1034 (m), 977 (m), 850 (w), 742 (m), 697 (m), 580 (vw), 489 (vw), 440 (vw). – **MS** (EI, 70 eV), m/z (%): 373/374 (25/26) [M]$^+$, 194 (100) [C$_{10}$H$_{12}$NO$_3$]$^+$, 135 (11) [C$_8$H$_7$O$_2$]$^+$, 134 (99)

$[C_8H_8NO]^+$, 122 (11) $[C_7H_8NO]^+$, 107 (14) $[C_7H_7O]^+$, 91 (48) $[C_6H_5N]^+$, 59 (13) $[C_2H_3O_2]^+$. – **HRMS** (EI, $C_{20}H_{20}O_4N^{35}Cl$) calc. 373.1081, found 373.1082.

6-Benzyl 1-methyl (*E*)-5-([1,1'-biphenyl]-2-ylamino)hex-2-enedioate (177fb)

General procedure 4a (GP 4a) was followed by stirring Py-box (**172**, 2.21 mg, 6.00 μmol, 6 mol%), NaBArF (5.20 mg, 6.00 μmol, 6 mol%) and Cu(MeCN)$_4$PF$_6$ (1.60 mg, 5.00 μmol, 5 mol%) in abs. CH$_2$Cl$_2$ (1 mL) in an oven dried vial at 40 °C for 2 h. 6-Benzyl 1-methyl (*E*)-5-diazohex-2-enedioate (**177f**, 27.4 mg, 0.100 mmol, 1.00 equiv.) and 2-aminobiphenyl (20.3 mg, 0.120 mmol, 1.20 equiv.) in 1 mL of CH$_2$Cl$_2$ was added subsequently and the mixture was stirred at room temperature for 2 h. The resulting mixture was purified directly *via* column chromatography (*c*-Hex/EtOAc = 8:1 → 4:1) to give the title product **177fb** as colorless oil, 15.2 mg, 36.0 μmol, 36%.

R_f = 0.30 (*c*-Hex/EtOAc = 4:1). – **^{1}H NMR** (400 MHz, MeOD) δ/ppm = 7.43–7.35 (m, 3H, 3 × CH^{Ar}), 7.35–7.28 (m, 5H, CH$_2$*Ph*), 7.25–7.23 (m, 2H, 2 × CH^{Ar}), 7.15 (ddd, J = 8.2, ^3J = 7.4, ^3J = 1.6 Hz, 1H, CH^{Ar}), 7.04 (dd, ^3J = 7.4, ^4J = 1.6 Hz, 1H, CH^{Ar}), 6.82–6.70 (m, 2H, CH^{Ar} + CH$_2$C*H*=), 6.68 (dd, ^3J = 8.2, ^4J = 1.1 Hz, 1H, CH^{Ar}), 5.73 (dt, $^3J_{trans}$ = 15.6, ^4J = 1.4 Hz, 1H, =C*H*CO$_2$Me), 5.18–5.04 (m, 2H, C*H$_2$*Ph), 4.37 (t, ^3J = 6.1 Hz, 1H, C*H*N), 3.69 (s, 3H, CO$_2$C*H$_3$*), 2.79–2.39 (m, 2H, C*H$_2$*CH=). – **^{13}C NMR** (101 MHz, MeOD) δ/ppm = 174.05 (C$_q$, CO$_2$Bn), 167.78 (C$_q$, CO$_2$CH$_3$), 144.65 (+, CH, CH$_2$CH=), 144.58 (C$_q$, *C*^{Ar}-NH), 140.53 (C$_q$, *C*^{Ar}), 137.03 (C$_q$, *C*^{Ar}), 131.36 (+, CH, *C*^{Ar}), 130.34 (+, CH, 2 × *C*^{Ar}), 130.22 (C$_q$, *C*^{Ar}), 129.92 (+, CH, 2 × *C*^{Ar}), 129.67 (+, CH, *C*^{Ar}), 129.57 (+, CH, 2 × *C*^{Ar}), 129.44 (+, CH, 2 × *C*^{Ar}), 129.37 (+, CH, *C*^{Ar}), 128.39 (+, CH, =CHCO$_2$Me), 125.31 (+, CH, *C*^{Ar}), 119.50 (+, CH, *C*^{Ar}), 112.88 (+, CH, *C*^{Ar}), 68.04 (–, *C*H$_2$Ph), 56.73 (+, CH, N*C*H), 52.03 (+, CH$_3$, CO$_2$*C*H$_3$), 35.81 (–, *C*H$_2$CH=).– **IR** (ATR): $\tilde{v}$/cm^{-1} = 2922 (w), 2852 (w), 2111 (vw), 1723 (m), 1658 (w), 1582 (w), 1511 (w), 1436 (w), 1377 (vw), 1268 (w), 1167 (m), 1008 (w), 803 (vw), 747 (w), 699 (m), 483 (vw).– **MS** (FAB, 3-NBA), m/z (%): 416 (12) [M + H]$^+$, 316 (17) [M – C$_2$H$_7$O$_2$]$^+$. – **HRMS** (FAB, C$_{26}$H$_{26}$O$_4$N$_1$, [M+H]$^+$): calc. 416.1862, found 416.1860.

6-Benzyl 1-methyl (*E*)-5-((3-methoxyphenyl)amino)hex-2-enedioate (177fd)

General procedure 4a (GP 4a) was followed by stirring Py-box (**172**, 2.21 mg, 6.00 μmol, 6 mol%), NaBArF (5.20 mg, 6.00 μmol, 6 mol%) and Cu(MeCN)$_4$PF$_6$ (1.60 mg, 5.00 μmol, 5 mol%) in abs. CH$_2$Cl$_2$ (1 mL) in an oven dried vial at 40 °C for 2 h. 6-Benzyl 1-methyl (*E*)-5-diazohex-2-enedioate (**103f**, 27.4 mg, 0.100 mmol, 1.00 equiv.) in 1 mL of CH$_2$Cl$_2$ and *m*-anisidine (15.0 mg, 0.120 mmol, 1.20 equiv.) were added subsequently and the mixture was stirred at room temperature for 2 h. The resulting mixture was purified directly *via* column chromatography (*c*-Hex/EtOAc = 8:1 → 6:1) to give the title product **177fd** as colorless oil, 15.2 mg, 36.0 μmol, 36%.

R_f = 0.18 (*c*-Hex/EtOAc = 8:1). – **^{1}H NMR** (400 MHz, MeOD) δ/ppm = 7.30–7.25 (m, 5H, CH$_2$*Ph*), 6.99 (t, ^3J = 8.0 Hz, 1H, C*H*^{Ar}), 6.95 (dt, $^3J_{trans}$ = 15.7, ^3J = 7.8 Hz, 1H, CH$_2$C*H*=), 6.28–6.20 (m, 3H, 3 × C*H*^{Ar}), 5.90 (dt, $^3J_{trans}$ = 15.7, ^4J = 1.3 Hz, 1H, =C*H*CO2Me), 5.13 (d, J = 2.4 Hz, 2H, C*H*$_2$Ph), 4.24 (t, ^3J = 6.8 Hz, 1H, C*H*N), 3.70 (s, 3H, OC*H*$_3$), 3.69 (s, 3H, CO$_2$C*H*$_3$), 2.88–2.41 (m, 2H, C*H*$_2$CH=). – **^{13}C NMR** (101 MHz, MeOD) δ/ppm = 174.62 (C$_q$, *C*O$_2$Bn), 168.14 (C$_q$, *C*O$_2$CH$_3$), 162.20 (C$_q$, *C*^{Ar}-OMe), 149.98 (C$_q$, *C*^{Ar}–NH), 145.5 (+, CH, CH$_2$*C*H=), 137.17 (C$_q$, *C*^{Ar}–CH$_2$), 130.87 (+, CH, *C*^{Ar}), 129.49 (+, CH, 2 × *C*^{Ar}), 129.36 (+, CH, 2 × *C*^{Ar}), 129.26 (+, CH, *C*^{Ar}), 124.62 (+, CH, =*C*HCO$_2$Me), 107.33 (+, CH, *C*^{Ar}), 104.62 (+, CH, *C*^{Ar}), 100.43 (+, CH, *C*^{Ar}), 67.83 (–, CH$_2$, *C*H$_2$Ph), 57.18 (+, CH, *C*HN), 55.40 (+, CH$_3$, O*C*H$_3$), 52.00 (+, CH$_3$, CO$_2$*C*H$_3$), 35.97 (–, CH$_2$, *C*H$_2$CH=). – **IR** (ATR): ṽ/cm^{-1} = 2961 (vw), 2117 (vw), 1724 (w), 1602 (vw), 1514 (vw), 1457 (vw), 1436 (vw), 1259 (w), 1170 (w), 1088 (w), 1020 (vw), 866 (vw), 797 (w), 743 (vw), 698 (vw), 399 (vw). – **MS** (EI, 70 eV), m/z (%): 369 (14) [M]$^+$, 270 (51) [M – C$_5$H$_7$O$_2$]$^+$, 234 (13) [M – C$_8$H$_7$O$_2$]$^+$, 91 (100) [C$_7$H$_7$]$^+$.– **HRMS** (EI, C$_{21}$H$_{23}$O$_5$Nl) calc. 369.1576, found 369.1577.

1-Benzyl 6-methyl (*E*)-5-(phenylamino)hex-2-enedioate (176d)

General procedure 4a (GP 4a) was followed by stirring Py-box (**172**, 2.21 mg, 6.00 μmol, 6 mol%), NaBArF (5.20 mg, 6.00 μmol, 6 mol%) and Cu(MeCN)$_4$PF$_6$ (1.60 mg, 5.00 μmol, 5 mol%) in abs. CH$_2$Cl$_2$ (1 mL) in an oven dried vial at 40 °C for 2 h. 1-Benzyl 6-methyl (*E*)-5-diazohex-2-enedioate (**103e**, 27.4 mg, 0.100 mmol, 1.00 equiv.) in 1 mL of CH$_2$Cl$_2$ and aniline (11.2 mg, 0.120 mmol, 1.20 equiv.) were added subsequently and the mixture was stirred at room temperature for 2 h. The resulting mixture was purified directly *via* column chromatography (*c*-Hex/EtOAc = 8:1 → 5:1) to give the title product **177d** as colorless oil, 20.4 mg, 60.0 μmol, 60%.

R_f = 0.37 (*c*-Hex/EtOAc =3:1). – **¹H NMR** (400 MHz, DMSO-d_6): δ/ppm = 7.38–7.32 (m, 5H, CH$_2$*Ph*), 7.08 (t, ^3J = 7.8 Hz, 2H, 2 × C*H*^{Ar}), 6.97 (dt, $^3J_{trans}$ = 15.6, ^3J = 7.2 Hz, 1H, CH$_2$C*H*=CH), 6.60–6.56 (m, 2H, 2 × C*H*^{Ar}), 6.07 (d, ^3J = 8.9 Hz, 1H, C*H*^{Ar}), 6.03 (dt, $^3J_{trans}$ = 15.6, ^4J = 1.6 Hz, 1H, CH$_2$CH=C*H*), 5.15 (s, 2H, C*H*$_2$Ph), 4.24 (t, ^3J = 5.6 Hz, 1H, C*H*N), 3.61 (s, 3H, OC*H*$_3$),2.78–2.58 (m, 2H, C*H*$_2$CH=). – **¹³C NMR** (101 MHz, DMSO-d_6): δ/ppm = 173.35 (C$_q$, *C*O$_2$Me), 165.17 (C$_q$, *C*O$_2$Bn), 147.37 (C$_q$, *C*^{Ar}-NH), 145.31 (+, CH, CH$_2$*C*H=), 136.09 (C$_q$, *C*^{Ar}-CH$_2$), 128.90 (+, CH, 2 × *C*^{Ar}), 128.41 (+, CH, 2 × *C*^{Ar}), 128.02 (+, CH, 3 × *C*^{Ar}), 123.02 (+, CH, =*C*HCO$_2$Bn), 116.68 (+, CH, *C*^{Ar}), 112.44 (+, CH, 2 × *C*^{Ar}), 65.40 (–, CH$_2$, *C*H$_2$Ph), 54.63 (+, CH, N*C*H), 51.77 (+, CH$_3$, O*C*H$_3$), 34.31 (–, CH$_2$, *C*H$_2$CH=).– **IR** (ATR): $\tilde{v}$/cm^{-1} = 3377 (vw), 3032 (vw), 2951 (vw), 1717 (m), 1655 (w), 1602 (w), 1498 (w), 1434 (w), 1377 (vw), 1312 (w), 1261 (w), 1211 (w), 1164 (m), 1081 (w), 979 (w), 749 (w), 679 (w), 521 (vw), 506 (vw). – **MS** (EI, 70 eV), m/z (%): 339 (12) [M]$^+$, 280 (11) [C$_{18}$H$_{18}$NO$_2$]$^+$, 164 (100) [C$_9$H$_9$NO$_2$]$^+$, 104 (45) [C$_7$H$_6$N]$^+$, 91 (72) [C$_6$H$_5$N]$^+$, 77 (20) [C$_6$H$_5$]$^+$. – **HRMS** (EI, C$_{20}$H$_{21}$O$_4$N) calc 339.1417, found 339.1469.

1-Ethyl 6-methyl (*E*)-5-(phenylamino)hex-2-enedioate (177b)

General procedure 4a (GP 4a) was followed by stirring Py-box (**172**, 2.21 mg, 6.00 μmol, 6 mol%), NaBArF (5.20 mg, 6.00 μmol, 6 mol%) and Cu(MeCN)$_4$PF$_6$ (1.60 mg, 5.00 μmol, 5 mol%) in abs. CH$_2$Cl$_2$ (1 mL) in an oven dried vial at 40 °C for 2 h. 1-Ethyl 6-methyl (*E*)-5-diazohex-2-enedioate (**103b**, 21.3 mg, 0.100 mmol, 1.00 equiv.) in 1 mL of CH$_2$Cl$_2$ and aniline (11.2 mg, 0.120 mmol, 1.20 equiv.) were added subsequently and the mixture was stirred at room temperature for 2 h. The resulting mixture was purified directly *via* column chromatography (*c*-Hex/EtOAc = 8:1 → 5:1) to give the title product **177b** as colorless oil, 13.2 mg, 49.0 μmol, 49%.

R_f = 0.17 (*c*-Hex/EtOAc = 5:1). – **¹H NMR** (400 MHz, DMSO-d_6) δ/ppm = 7.11–7.03 (m, 2H, 2 × C*H*^{Ar}), 6.91 (dt, $^3J_{trans}$ = 15.7, ^3J = 7.1 Hz, 1H, CH$_2$C*H*=), 6.61–6.54 (m, 3H, 3 × C*H*^{Ar}), 6.07 (d, ^3J = 9.0 Hz, 1H, C*H*^{Ar}), 5.96 (dt, $^3J_{trans}$ = 15.5, ^4J = 1.5 Hz, 1H, =C*H*CO$_2$Et), 4.22 (t, ^3J = 5.6 Hz, 1H, C*H*N), 4.10 (q, ^3J = 7.1 Hz, 2H, OC*H*$_2$CH$_3$), 3.62 (s, 3H, OC*H*$_3$), 2.76–2.55 (m, 2H, C*H*$_2$CH=), 1.20 (t, ^3J = 7.1 Hz, 3H, OCH$_2$C*H*$_3$). – **¹³C NMR** (101 MHz, DMSO-d_6) δ/ppm = 173.36 (C$_q$, *C*O$_2$Me), 165.32 (C$_q$, *C*O$_2$Et), 147.37 (C$_q$, *C*^{Ar}–NH), 144.63 (+, CH, CH$_2$*C*H=), 128.90 (+, CH, 2 × *C*^{Ar}), 123.30 (+, CH, =*C*HCO$_2$Et),

116.68 (+, CH, C^{Ar}), 112.43 (+, CH, 2 × C^{Ar}), 59.81 (–, CH_2, OCH_2CH_3), 54.65 (+, CH, NCH), 51.78 (+, CH_3, CO_2CH_3), 34.23 (–, CH_2, $CH_2CH=$), 14.09 (+, CH_3, OCH_2CH_3).

1-Ethyl 6-methyl (*E*)-5-(p-tolylamino)hex-2-enedioate (177bl)

General procedure 4a (GP 4a) was followed by stirring Py-box (**172**, 2.21 mg, 6.00 μmol, 6 mol%), NaBArF (5.20 mg, 6.00 μmol, 6 mol%) and Cu(MeCN)$_4$PF$_6$ (1.60 mg, 5.00 μmol, 5 mol%) in abs. CH$_2$Cl$_2$ (1 mL) in an oven dried vial at 40 °C for 2 h. 1-Ethyl 6-methyl (*E*)-5-diazohex-2-enedioate (**177b**, 21.3 mg, 0.100 mmol, 1.00 equiv.) in 1 mL of CH$_2$Cl$_2$ and *p*-toluidine (12.9 mg, 0.120 mmol, 1.20 equiv.) were added subsequently and the mixture was stirred at room temperature for 2 h. The resulting mixture was purified directly *via* column chromatography (*c*-Hex/EtOAc = 8:1 → 5:1) to give the title product **177bl** as colorless oil, 13.2 mg, 45.0 μmol, 45%.

R_f = 0.33 (*c*-Hex/EtOAc = 3:1). – **H NMR** (400 MHz, DMSO-d_6) δ/ppm = 6.90–6.74 (m, 3H, 2 × CH^{Ar} + $CH_2CH=$), 6.42 (d, ^3J = 8.4 Hz, 2H, 2 × CH^{Ar}), 5.88 (dt, $^3J_{trans}$ = 15.5, ^4J = 1.5 Hz, 1H, $=CHCO_2Et$), 5.78 (d, ^3J = 9.2 Hz, 1H, CH^{Ar}), 4.15–4.07 (m, 1H, CHN), 4.03 (q, ^3J = 7.1 Hz, 2H, OCH_2CH_3), 3.53 (s, 3H, OCH_3), 2.65–2.51 (m, 2H, $CH_2CH=$), 2.07 (s, 3H, CH_3), 1.13 (t, ^3J = 7.1 Hz, 3H, OCH_2CH_3). – **^{13}C NMR** (101 MHz, DMSO-d_6) δ/ppm = 173.49 (C_q, CO_2Me), 165.33 (C_q, CO_2Et), 145.07 (+, CH, $CH_2CH=$), 144.71 (C_q, C^{Ar}–NH), 129.33 (+, CH, 2 × C^{Ar}), 125.15 (C_q, C^{Ar}), 123.24 (+, CH, $=CHCO_2Et$), 112.61 (+, CH, 2 × C^{Ar}), 59.80 (–, CH_2, OCH_2CH_3), 54.96 (+, CH, CHN), 51.71 (+, CH, CO_2CH_3), 34.28 (–, CH_2, $CH_2CH=$), 20.00 (+, CH_3, CH_3), 14.09 (+, CH_3, OCH_2CH_3).

1-*tert*-Butyl 6-methyl (*E*)-5-(phenylamino)hex-2-enedioate (176c)

General procedure 4a (GP 4a) was followed by stirring Py-box (**172**, 2.21 mg, 6.00 μmol, 6 mol%), NaBArF (5.20 mg, 6.00 μmol, 6 mol%) and Cu(MeCN)$_4$PF$_6$ (1.60 mg, 5.00 μmol, 5 mol%) in abs. CH$_2$Cl$_2$ (1 mL) in an oven dried vial at 40 °C for 2 h. 1-*tert*-Butyl 6-methyl (*E*)-5-diazohex-2-enedioate (**103c**, 24.0 mg, 0.100 mmol, 1.00 equiv.) in 1 mL of CH$_2$Cl$_2$ and aniline (11.2 mg, 0.120 mmol, 1.20 equiv.) were added subsequently and the mixture was stirred at room temperature for 2 h. The resulting mixture was purified directly *via* column chromatography (*c*-Hex/EtOAc = 5:1) to give the title product **177c** as yellow liquid, 13.5 mg, 44.0 μmol, 39%.

R_f = 0.19 (*c*-Hex/EtOAc = 4:1). – **^{1}H NMR** (500 MHz, MeOD) δ/ppm = 7.14–7.07 (m, 2H, 2 × C*H*^{Ar}), 6.86 (dt, $^3J_{trans}$ = 15.7, ^3J = 7.3 Hz, 1H, CH$_2$C*H*=), 6.68–6.59 (m, 3H, 3 × C*H*^{Ar}), 5.85 (dt, $^3J_{trans}$ = 15.6, ^4J = 1.6 Hz, 1H, =C*H*CO$_2$*t*Bu), 4.26–4.12 (m, 1H, C*H*N), 3.69 (s, 3H, CO$_2$C*H$_3$*), 2.79–2.44 (m, 2H, C*H$_2$*CH=), 1.46 (s, 9H,$^{\circ}$C(C*H$_3$*)$_3$). – **^{13}C NMR** (126 MHz, MeOD) δ/ppm = 172.44 (C$_q$, *C*O$_2$Me), 164.32 (C$_q$, *C*O$_2$*t*Bu), 145.67 (C$_q$, *C*^{Ar}), 141.40 (+, CH, CH$_2$*C*H=), 127.20 (+, CH, 2 × *C*^{Ar}), 123.68 (+, CH, =*C*HCO$_2$*t*Bu), 116.05 (+, CH, *C*^{Ar}), 111.49 (+, CH, 2 × *C*^{Ar}), 78.76 (C$_q$, *C*(CH$_3$)$_3$), 54.13 (+, CH, N*C*H), 49.63 (+, CH$_3$, CO$_2$*C*H$_3$), 33.03 (–, CH$_2$, *C*H$_2$CH=), 25.43 (+, CH$_3$,$^{\circ}$C(*C*H$_3$)$_3$). – **IR** (ATR): $\tilde{v}$/cm^{-1} = 3344 (vw), 2923 (w), 2479 (vw), 1726 (m), 1697 (m), 1658 (w), 1601 (m),1528 (vw), 1498 (w), 1438 (w),1368 (w),1350 (w), 1313 (w), 1255 (m), 1234 (m), 1211 (m), 1145 (m), 1099 (w), 1051 (w), 981 (m), 907 (vw), 875 (vw), 845 (w), 777 (w), 753 (m), 693 (m), 598 (vw), 536 (w), 465 (w). – **MS** (EI, 70 eV), m/z (%): 305 (8) [M]$^+$, 204 (87) [M – C$_5$H$_9$O$_2$]$^+$, 172 (33) [M – C$_5$H$_9$O$_2$ – CH$_3$O – H]$^+$, 164 (86) [C$_9$H$_{10}$NO$_2$]$^+$, 165 (100) [C$_9$H$_{10}$NO$_2$ + H]$^+$, 104 (56) [C$_7$H$_7$N – H]$^+$. – **HRMS** (EI, C$_{17}$H$_{23}$O$_4$N) calc. 305.1622, found 305.1623.

6-*tert*-Butyl 1-methyl (*E*)-5-(phenylamino)hex-2-enedioate (176f)

General procedure 4a (GP 4a) was followed by stirring Py-box (**172**, 2.21 mg, 06.00 μmol, 6 mol%), NaBArF (5.20 mg, 6.00 μmol, 6 mol%) and Cu(MeCN)$_4$PF$_6$ (1.60 mg, 5.00 μmol, 5 mol%) in abs. CH$_2$Cl$_2$ (1 mL) in an oven dried vial at 40 °C for 2 h. 1-Methyl 6-*tert*-butyl (*E*)-5-diazohex-2-enedioate (**103d**, 24.0 mg, 0.100 mmol, 1.00 equiv.) in 1 mL of CH$_2$Cl$_2$ and aniline (11.2 mg, 0.120 mmol, 1.20 equiv.) were added subsequently and the mixture was stirred at room temperature for 2 h. The resulting mixture was purified directly *via* column chromatography (*c*-Hex/EtOAc = 5:1) to give the title product **176f** as yellow liquid, 27.4 mg, 75.0 μmol, 75%.

R_f = 0.20 (*c*-Hex/EtOAc = 4:1). – **^{1}H NMR** (400 MHz, MeOD) δ/ppm = 7.05–6.97 (m, 2H, 2 × C*H*^{Ar}), 6.89 (dt, $^3J_{trans}$ = 15.7, ^3J = 7.3 Hz, 1H, CH$_2$C*H*=), 6.61–6.50 (m, 3H, 3 × C*H*^{Ar}), 5.86 (dt, $^3J_{trans}$ = 15.6, ^4J = 1.5 Hz, 1H, =C*H*CO$_2$M*e*), 3.98 (t, ^3J = 6.8 Hz, 1H, C*H*N), 3.61 (s, 3H, C*H$_3$*), 2.58 (td, ^3J = 7.2, ^4J = 1.3 Hz, 2H, C*H$_2$*CH=), 1.30 (s, 9H,$^{\circ}$C(C*H$_3$*)$_3$). – **^{13}C NMR** (101 MHz, MeOD) δ/ppm = 174.1 (C$_q$, *C*O$_2$*t*Bu), 168.21 (C$_q$, *C*O$_2$Me), 148.68 (C$_q$, *C*^{Ar}-NH), 145.85 (+, CH, CH$_2$*C*H=), 130.05 (+, CH, 2 × *C*^{Ar}), 124.50 (+, CH, =*C*HCO$_2$Me), 119.0 (+, CH, *C*$_{Ar}$), 114.66 (–, CH, 2 × *C*^{Ar}), 82.94 (C$_q$, *C*(CH$_3$)$_3$), 57.79 (+, CH, N*C*H), 52.03 (+, CH$_3$, CO$_2$*C*H$_3$), 36.10 (–, CH, *C*H$_2$CH=), 28.25 (+, CH$_3$,$^{\circ}$C(*C*H$_3$)$_3$). – **IR** (ATR): $\tilde{v}$/cm^{-1} = 3354 (m), 2983 (w), 1727 (s), 1702 (s), 1659 (m),

1600 (m),1518 (w), 1494 (m), 1433 (m),1365 (m), 1344 (m), 1315 (m), 1263 (s), 1220 (s), 1150 (s), 1082 (m), 1036 (m), 986 (m), 981 (m), 848 (m), 845 (w), 780 (m), 746 (s), 718 (w), 690 (m), 582 (w), 540 (w), 489 (m), 417 (w). – **MS** (EI, 70 eV), m/z (%): 305 (22) [M]$^+$, 204 (87) [M – C$_5$H$_9$O$_2$]$^+$, 172 (33) [M – C$_5$H$_9$O$_2$ – CH$_3$O – H]$^+$, 150 (100) [C$_8$H$_7$NO$_2$ + H]$^+$, 104 (44) [C$_7$H$_7$N – H]$^+$. – **HRMS** (EI, C$_{17}$H$_{23}$O$_4$N) calc. 305.1627, found 305.1626.

6-Benzyl 1-methyl (*E*)-5-((3, 5-dimethoxyphenyl)amino)hex-2-enedioate (177fi)

General procedure 4a (GP 4a) was followed by stirring Py-box (**172**, 2.21 mg, 6.00 μmol, 6 mol%), NaBArF (5.20 mg, 6.00 μmol, 6 mol%) and Cu(MeCN)$_4$PF$_6$ (1.6 mg, 5.00 μmol, 5 mol%) in abs. CH$_2$Cl$_2$ (1 mL) in an oven dried vial at 40 °C for 2 h. 6-Benzyl 1-methyl (*E*)-5-diazohex-2-enedioate (**103f**, 27.4 mg, 0.100 mmol, 1.00 equiv.) and 3,5-dimethoxyaniline (18.4 mg, 0.120 mmol, 1.20 equiv.) in 1 mL of CH$_2$Cl$_2$ was added subsequently and the mixture was stirred at room temperature for 2 h. The crude was purified *via* flash chromatography (*c*-Hex/EtOAc = 6:1 → 4:1) to give product **177fi** as colorless solid, 6.70 mg, 17.0 μmol, 17%.

R_f = 0.13 (*c*-Hex/EtOAc = 4:1). – **^{1}H NMR** (500 MHz, CDCl$_3$) δ/ppm = 7.37–7.28 (m, 5H, CH$_2$*Ph*), 6.87 (dt, $^3J_\text{trans}$ = 15.2, ^3J = 7.4 Hz, 1H, CH$_2$C*H*=), 5.93 (s, 1H, C*H*$^\text{Ar}$), 5.86 (dt, $^3J_\text{trans}$ = 15.7, ^4J = 1.5 Hz, 1H, =C*H*CO$_2$Me), 5.81 (d, ^4J = 1.9 Hz, 2H, 2 × C*H*$^\text{Ar}$), 5.17 (d, J = 2.8 Hz, 2H, C*H*$_2$Ph), 4.25 (t, ^3J = 6.1 Hz, 1H, C*H*N), 3.72 (s, 6H, 2 × OC*H*$_3$), 3.71 (s, 3H, C*H*$_3$), 2.82–2.56 (m, 2H, C*H*$_2$CH=). – **^{13}C NMR** (126 MHz, CDCl$_3$) δ/ppm = 172.31 (C$_q$, *C*O$_2$Bn), 166.18 (C$_q$, *C*O$_2$CH$_3$), 161.79 (C$_q$, 2 × *C*$^\text{Ar}$–OMe), 147.74 (C$_q$, *C*$^\text{Ar}$-NH), 142.63 (+, CH, CH$_2$*C*H=), 135.12 (C$_q$, *C*$^\text{Ar}$-CH$_2$), 128.65 (+, CH, 2 × *C*$^\text{Ar}$), 128.56 (+, CH, *C*$^\text{Ar}$), 128.51 (+, CH, 2 × *C*$^\text{Ar}$), 124.52 (+, CH, =*C*HCO$_2$Me), 92.61 (+, CH, 2 × *C*$^\text{Ar}$), 91.33 (+, CH, *C*$^\text{Ar}$), 67.36 (–, CH$_2$, *C*H$_2$Ph), 55.68 (+, CH, NH*C*H), 55.20 (+, CH$_3$, 2 × O*C*H$_3$), 51.61 (+, CH$_3$, CO$_2$*C*H$_3$), 35.11 (–, CH$_2$, *C*H$_2$CH=). – **IR** (ATR): ṽ/cm^{-1} = 3377 (vw), 2950 (w), 2841 (vw), 1720 (m), 1658 (w), 1596 (v), 1518 (w), 1482 (w), 1455 (m), 1435 (m), 1379 (w), 1319 (w), 1272 (m), 1202 (s), 1151 (s), 1066 (m), 975 (m), 911 (w), 813 (m), 734 (m), 698 (m), 538 (vw). – **MS** (EI, 70 eV), m/z (%): 399 (1) [M]$^+$, 91 (16) [C$_7$H$_7$]$^+$, 59 (11) [C$_2$H$_3$O$_2$]$^+$. – **HRMS** (EI, C$_{22}$H$_{25}$O$_6$N$_1$) calc. 399.1682, found 399.1681.

1-Benzyl 6-methyl (*E*)-5-((4-((trimethylsilyl)ethynyl)phenyl)amino)hex-2-enedioate (177fj)

General procedure 4b (GP 4b) was followed by stirring Ru-Complex (3.80 mg, 6.00 μmol, 6 mol%) and NaBArF (5.20 mg, 6.00 μmol, 6 mol%) in abs. CH$_2$Cl$_2$ (1 mL) in an oven dried vial. 6-Benzyl 1-methyl (*E*)-5-diazohex-2-enedioate (**177f**, 27.4 mg, 0.100 mmol, 1.00 equiv.) and 4-((trimethylsilyl)ethynyl)aniline (22.7 mg, 0.120 mmol, 1.20 equiv.) in 1 mL of CH$_2$Cl$_2$ was added subsequently and the mixture was stirred at room temperature for 2 h. The resulting mixture was purified directly *via* column chromatography (pentane/Et$_2$O = 5:1) to give the title product **177fj** as colorless solid, 30.4 mg, 70.0 μmol, 70%.

R_f = 0.21 (pentane/Et$_2$O = 4:1). – **^{1}H NMR** (400 MHz, CDCl$_3$) δ/ppm = 7.40–7.32 (m, 5H, CH$_2$*Ph*), 7.31–7.28 (m, 2H, 2 × C*H*^{Ar}), 6.92 (dt, $^3J_{trans}$ = 14.8, ^3J = 7.4 Hz, 1H, CH$_2$C*H*=), 6.54–6.47 (m, 2H, 2 × C*H*^{Ar}), 5.95 (dt, $^3J_{trans}$ = 15.5, ^4J = 1.5 Hz, 1H, CH$_2$CH=C*H*), 5.17 (s, 2H, C*H*$_2$Ph), 4.25 (q, ^3J = 6.1 Hz, 1H, C*H*N), 3.73 (s, 3H, OC*H*$_3$), 2.91–2.52 (m, 2H, C*H*$_2$CH=CH), 0.22 (s, 9H, SiC(C*H*$_3$)$_3$). – **^{13}C NMR** (101 MHz, CDCl$_3$) δ/ppm = 172.49 (C$_q$, *C*O$_2$Me), 165.41 (C$_q$, *C*O$_2$Bn), 145.95 (C$_q$, *C*^{Ar}-NH), 142.75 (+, CH, CH$_2$*C*H=CH), 135.67 (C$_q$, *C*^{Ar}–CH$_2$), 133.31 (+, CH, 2 × *C*^{Ar}), 128.44 (+, CH, 2 × *C*^{Ar}), 128.15 (+, CH, *C*^{Ar}), 128.13 (+, CH, 2 × *C*^{Ar}), 124.53 (+, CH, CH$_2$CH=*C*H), 112.84 (+, CH, 2 × *C*^{Ar}), 112.59 (C$_q$, *C*^{Ar}–C≡C), 105.63 (C$_q$, *C*≡CTMS), 91.62 (C$_q$, C≡*C*TMS), 66.21 (–, CH$_2$, *C*H$_2$Ph), 54.97 (+, CH, N*C*H), 52.42 (+, CH$_3$, O*C*H$_3$), 35.02 (–, CH$_2$, *C*H$_2$CH=CH), 0.00 (+, CH$_3$, Si(*C*H$_3$)$_3$). – **IR** (ATR): $\tilde{v}$/cm^{-1} = 3380 (vw), 2955 (w), 2148 (m), 1718 (m), 1656 (w), 1607 (m), 1516 (m), 1437 (w), 1377 (w), 1320 (m), 1248 (m), 1211 (m), 1164 (m), 1014 (w), 980 (w), 862 (s), 839 (s), 757 (m), 696 (m), 638 (w), 537 (w), 455 (vw). – **MS** (FAB, 3-NBA), m/z (%): 436 (28) [M + H]$^+$, 376 (8) [M – CO$_2$Me]$^+$, 260 (91) [M – C$_{11}$H$_{11}$O$_2$]$^+$, 200 (56) [M – C$_{11}$H$_{11}$O$_2$ – CO$_2$Me]$^+$, 132 (100) [C$_8$H$_8$Si]$^+$. – **HRMS** (FAB, C$_{25}$H$_{30}$O$_4$N$_1$^{28}Si$_1$, [M+H]$^+$): calc. 436.1944, found 436.1945.

6-Benzyl 1-methyl (*E*)-5-((3,5-bis(trifluoromethyl)phenyl)amino)hex-2-enedioate (177fg)

General procedure 4a (GP 4a) was followed by stirring Py-box (**172**, 2.21 mg, 6.00 μmol, 6 mol%), NaBArF (5.20 mg, 6.00 μmol, 6 mol%) and Cu(MeCN)$_4$PF$_6$ (1.6 mg, 5.00 μmol, 5 mol%) in abs. CH$_2$Cl$_2$ (1 mL) in an oven dried vial at −0 °C for 2 h. 6-Benzyl 1-methyl (*E*)-5-diazohex-2-enedioate (**103f**, 27.4 mg, 0.100 mmol, 1.00 equiv.) in 1 mL of CH$_2$Cl$_2$ and

3,5-bis(trifluoromethyl)aniline (18.7 μL, 27.5 mg, 0.120 mmol, 1.20 equiv.) were added subsequently and the mixture was stirred at room temperature for 2 h. The resulting mixture was dried under vacuum and purified directly *via* column chromatography (pentane/Et$_2$O = 5:1) to give the title product **177fg** as colorless oil, 19.0 mg, 40.0 μmol, 40%.

R_f = 0.28 (*c*-Hex/EtOAc = 5:1). – **^{1}H NMR** (400 MHz, CDCl$_3$) δ/ppm = 7.40–7.28 (m, 5H, CH$_2$*Ph*), 7.22 (s, 1H, C*H*^{Ar}), 6.95 (s, 2H, 2 × C*H*^{Ar}), 6.84 (dt, $^3J_{trans}$ = 15.2, ^3J = 7.4 Hz, 1H, CH$_2$C*H*=), 5.89 (dt, $^3J_{trans}$ = 15.7, ^4J = 1.4 Hz, 1H, =C*H*CO$_2$Me), 5.19 (s, 2H, C*H*$_2$Ph), 4.32 (t, ^3J = 5.9 Hz, 1H, C*H*N), 3.73 (s, 3H, OC*H*$_3$), 2.92–2.60 (m, 2H, C*H*$_2$CH=). – **^{13}C NMR** (126 MHz, CDCl$_3$) δ/ppm = 171.53 (C$_q$, *C*O$_2$Bn), 166.10 (C$_q$, *C*O$_2$Me), 146.90 (C$_q$, *C*^{Ar}–NH), 141.75 (+, CH, CH$_2$*C*H=), 134.79 (C$_q$, *C*^{Ar}–CH$_2$), 128.95 (C$_q$, 2 × *C*^{Ar}–CF$_3$), 128.93 (C$_q$, 2 × *C*F$_3$), 128.87 (+, CH, 2 × *C*^{Ar}), 128.79 (+, CH, *C*^{Ar}), 128.72 (+, CH, 2 × *C*^{Ar}), 125.77 (+, CH, =*C*HCO$_2$Bn), 112.82 (+, CH, 2 × *C*^{Ar}), 111.89 (+, CH, *C*^{Ar}), 67.98 (–, CH$_2$, *C*H$_2$Ph), 55.12 (+, CH, *C*HN), 51.67 (+, CH$_3$, O*C*H$_3$), 34.75 (–, CH$_2$, *C*H$_2$CH=). – **IR** (ATR): $\tilde{v}$/cm^{-1} = 3335 (w), 2921 (w), 2116 (vw), 1736 (m), 1703 (m), 1661 (w), 1622 (w), 1550 (w), 1475 (w), 1441 (w), 1396 (m), 1352 (w), 1316 (w), 1276 (s), 1234 (m), 1213 (m), 1164 (s), 1119 (s), 1047 (m), 984 (m), 891 (w), 855 (m), 753 (w), 733 (m), 697 (m), 682 (m), 597 (w), 414 (w). – **MS** (FAB, 3-NBA), m/z (%): 476 (28) [M + H]$^+$, 376 (13) [M – C$_5$H$_7$O$_2$]$^+$, 340 (17) [M – C$_8$H$_7$O$_2$]$^+$. – **HRMS** (FAB, C$_{22}$H$_{20}$O$_4$N$_1$F$_6$, [M+H]$^+$): calc. 476.1297, found 476.1297.

6-Benzyl 1-methyl (*E*)-5-((3,5-di-*tert*-butylphenyl)amino)hex-2-enedioate (177fk)

General procedure 4a (GP 4a) was followed by stirring Py-box (**172**, 2.21 mg, 6.00 μmol, 6 mol%), NaBArF (5.20 mg, 6.00 μmol, 6 mol%) and Cu(MeCN)$_4$PF$_6$ (1.6 mg, 5.00 μmol, 5 mol%) in abs. CH$_2$Cl$_2$ (1 mL) in an oven dried vial at 40 °C for 2 h. 6-Benzyl 1-methyl (*E*)-5-diazohex-2-enedioate (**177f**, 27.4 mg, 0.100 mmol, 1.00 equiv.) and 3,5-di-*tert*butylaniline (24.6 mg, 0.120 mmol, 1.20 equiv.) in 1 mL of CH$_2$Cl$_2$ was added dropwise and the mixture was stirred at room temperature for 2 h. The resulting mixture was dried under vacuum and purified directly *via* column chromatography (pentane/Et$_2$O = 5:1) to give the title product **177fk** as colorless oil, 15.4 mg, 34.0 μmol, 34%.

R_f = 0.14 (pentane/Et$_2$O = 5:1). – **^{1}H NMR** (400 MHz, CDCl$_3$) δ/ppm = 7.40–7.27 (m, 6H, CH$_2$*Ph* + C*H*^{Ar}), 7.00–6.87 (m, 2H, CH$_2$C*H*=CH + C*H*^{Ar}), 6.47 (d, ^4J = 1.7 Hz, 2H, 2 × C*H*^{Ar}), 5.89 (dt, $^3J_{trans}$ = 15.5, ^4J = 1.5 Hz, 1H, CH$_2$CH=C*H*), 5.15 (d, J = 7.1 Hz, 2H, C*H*$_2$Ph), 4.30 (t, ^3J = 6.3 Hz, 1H, C*H*N), 3.72 (s, 3H, OC*H*$_3$), 2.73 (dq, ^2J = 14.1,

^3J = 7.4 Hz, 2H, CH_2CH=CH), 1.26 (s, 18H, 2 × C(CH_3)$_3$). – ^{13}**C NMR** (101 MHz, CDCl$_3$) δ/ppm = 173.78 (C$_q$, CO_2Bn), 167.14 (C$_q$, CO_2Me), 152.83 (C$_q$, 2 × C^{Ar}–tBu), 146.35 (C$_q$, C^{Ar}–NH), 144.02 (+, CH, CH$_2$CH=CH), 136.08 (C$_c$, C^{Ar}-CH$_2$), 129.52 (+, CH, 2 × C^{Ar}), 129.37 (+, CH, C^{Ar}), 129.30 (+, CH, 2 × C^{Ar}), 125.22 (+, CH, CH$_2$CH=CH), 114.38 (+, CH, C^{Ar}), 109.19 (+, CH, 2 × C^{Ar}), 68.10 (–, CH$_2$, CH_2Ph), 56.90 (+, CH, CHN), 52.46 (+, CH$_3$, OCH_3), 36.52 (–, CH$_2$, CH_2CH=CH), 35.73 (C$_q$, 2 × C(CH$_3$)$_3$), 32.19 (+, CH$_3$, 2 × C(CH_3)$_3$). – **IR** (ATR): $\tilde{v}$/cm^{-1} = 3377 (vw), 2954 (m), 2867 (w), 1724 (m), 1659 (w), 1598 (m), 1498 (vw), 1435 (m), 1393 (w), 1362 (w), 1247 (m), 1202 (m), 1165 (m), 1028 (w), 980 (w), 900 (w), 850 (w), 803 (vw), 735 (w), 697 (m), 580 (vw), 491 (vw), 396 (vw). – **MS** (FAB, 3-NBA), m/z (%): 452 (23) [M + H]$^+$, 352 (60) [M – C$_5$H$_7$O$_2$]$^+$, 316 (27) [M – C$_8$H$_7$O$_2$]$^+$, 216 (13) [C$_{15}$H$_{23}$N – H]$^+$, 133 (20) [C$_9$H$_{11}$N]$^+$. – **HRMS** (FAB, C$_{28}$H$_{38}$O$_4$N$_1$, [M+H]$^+$): calc. 452.2801, found 452.2800.

1-Benzyl 6-methyl (*E*)-5-((3,5-di-tert-butylphenyl)amino)hex-2-enedioate (177ek)

General procedure 4a (GP 4a) was followed by stirring Py-box (**172**, 2.21 mg, 6.00 µmol, 6 mol%), NaBArF (5.20 mg, 6.00 µmol, 6 mol%) and Cu(MeCN)$_4$PF$_6$ (1.60 mg, 5.00 µmol, 5 mol%) in abs. CH$_2$Cl$_2$ (1 mL) in an oven dried vial at 40 °C for 2 h. 1-Benzyl 6-methyl (*E*)-5-diazohex-2-enedioate (**177e**, 27.4 mg, 0.100 mmol, 1.00 equiv.) and 3,5-di-*tert*butylaniline (24.6 mg, 0.120 mmol, 1.20 equiv.) in 1 mL of CH$_2$Cl$_2$ was added dropwise and the mixture was stirred at room temperature for 2 h. The resulting mixture was dried under vacuum and purified directly *via* column chromatography (pentane/Et$_2$O = 5:1) to give the title product **177ek** as colorless oil, 18.7 mg, 41.0 µmol, 41%.

R_f = 0.13 (pentane/Et$_2$O = 3:1). – ^{1}**H NMR** (500 MHz, CDCl$_3$) δ/ppm = 7.45–7.31 (m, 5H, CH$_2$Ph), 6.99 (dt, $^3J_{trans}$ = 15.1, ^3J = 7.3 Hz, 1H, CH$_2$CH=CH), 6.85 (t, ^4J = 1.6 Hz, 1H, CH^{Ar}), 6.47 (d, J = 1.6 Hz, 2H, CH^{Ar}), 5.99 (dt, $^3J_{trans}$ = 15.6, ^4J = 1.4 Hz, 1H, CH$_2$CH=CH), 5.18 (s, 2H, CH_2Ph), 4.26 (t, ^3J = 6.2 Hz, 1H, CHN), 3.74 (s, 3H, CO$_2$CH_3), 2.90–2.57 (m, 2H, CH_2CH=CH), 1.28 (s, 18H, 2 × C(CH_3)$_3$). – ^{13}**C NMR** (126MHz, CDCl$_3$) δ/ppm = 173.65 (C$_q$, CO_2Me), 165.82 (C$_q$, CO_2Bn), 152.01 (C$_q$, 2 × C^{Ar}–tBu), 145.47 (C$_q$, C^{Ar}–NH), 143.75 (+, CH, CH$_2$CH=CH), 135.97 (C$_q$, C^{Ar}–CH$_2$), 128.69 (+, CH, 2 × C^{Ar}), 128.41 (+, CH, 2 × C^{Ar}), 128.39 (+, CH, C^{Ar}), 124.48 (+, CH, CH$_2$CH=CH), 113.54 (+, CH, C^{Ar}), 108.34 (+, CH, 2 × C^{Ar}), 66.42 (–, CH$_2$, CH_2Ph), 55.93 (+, CH, CHN), 52.52 (+, CH$_3$, OCH_3), 35.74 (–, CH$_2$, CH_2CH=CH), 34.97 (C$_q$, 2 × C(CH$_3$)$_3$), 31.53 (+, CH$_3$, 2 × C(CH_3)$_3$). – **IR** (ATR): $\tilde{v}$/cm^{-1} = 2954 (w), 2150 (vw), 1722 (m), 1655 (w), 1599 (m), 1490 (w), 1437 (w), 1371 (w), 1247 (m), 1213 (m), 1159 (m), 1027 (m), 849 (m), 754 (w), 731 (w), 697 (w), 664 (m), 639 (m), 613 (m), 523 (vw), 481 (vw), 411 (vw).

– **MS** (FAB, 3-NBA), m/z (%): 452 (24) [M + H]$^+$, 392 (8) [M – CO$_2$Me]$^+$, 276 (100) [M – C$_{11}$H$_{11}$O$_2$]$^+$, 216 (27) [C$_{15}$H$_{23}$N – H]$^+$, 133 (27) [C$_9$H$_{11}$N]$^+$. – **HRMS** (FAB, C$_{28}$H$_{38}$O$_4$N$_1$ [M+H]$^+$): calc. 452.2801, found 452.2800.

1-Benzyl 6-methyl (*E*)-5-((*rac*)-4-[2.2]paracyclophanylamino)hex-2-enedioate (178)

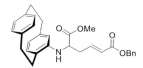

General procedure 4a (GP 4a) was followed by stirring Pybox (**172**, 2.21 mg, 6.00 μmol, 6 mol%), NaBArF (5.20 mg, 6.00 μmol, 6 mol%) and Cu(MeCN)$_4$PF$_6$ (1.60 mg, 5.00 μmol, 5 mol%) in 1 mL of abs. CH$_2$Cl$_2$ in an oven dried vial at 40 °C for 2 h. 1-Benzyl 6-methyl (*E*)-5-diazohex-2-enedioate (**177e**, 27.4 mg, 0.100 mmol, 1.00 equiv.) and (*rac*)-4-amino[2.2]paracyclophane (**161a**, 26.0 mg, 0.120 mmol, 1.20 equiv.) in CH$_2$Cl$_2$ (1 mL) was added dropwise, the mixture was stirred at room temperature overnight. The resulting mixture was dried under vacuum and purified directly *via* column chromatography (*c*-Hex/EtOAc = 5:1) to give the title product **178** as colorless oil, 7.10 mg, 14.0 μmol, 14%.

R_f = 0.32 (*c*-Hex/EtOAc = 8:1). – **^{1}H NMR** (500 MHz, CDCl$_3$) δ/ppm = 7.42–7.29 (m, 5H, CH$_2$*Ph*), 7.20 (ddd, *J* = 15.6, 8.1, 6.8 Hz, 1H, C*H*^{Ar}), 6.89 (dd, *J* = 7.8, 1.8 Hz, 1H, CH$_2$C*H*=CH), 6.55 (dd, *J* = 7.8, 1.8 Hz, 1H, C*H*^{Ar}), 6.35 (ddd, *J* = 14.1, 7.8, 1.8 Hz, 2H, C*H*^{Ar}), 6.29 (d, *J* = 7.6 Hz, 1H, C*H*^{Ar}), 6.19–6.06 (m, 2H, C*H*^{Ar} + CH$_2$CH=C*H*), 5.28 (d, *J* = 1.6 Hz, 1H, C*H*^{Ar}), 5.22 (s, 2H, C*H*$_2$Ph), 4.07 (brs, 1H, N*H*), 4.00 (t, ^{3}*J* = 6.6 Hz, 1H, C*H*N), 3.65 (s, 3H, OC*H*$_3$), 3.14 (ddd, *J* = 13.9, 8.2, 2.5 Hz, 1H, C*H*^{PC}), 3.07 – 2.91 (m, 5H, 5 × C*H*^{PC}), 2.91–2.79 (m, 2H, C*H*$_2$CH=CH), 2.79–2.72 (m, 1H, C*H*^{PC}), 2.68 (ddd, *J* = 13.8, 9.9, 7.5 Hz, 1H, C*H*^{PC}). – **^{13}C NMR** (126 MHz, CDCl$_3$) δ/ppm = 173.67 (C$_q$, *C*O$_2$Me), 165.73 (C$_q$, *C*O$_2$Bn), 144.85 (C$_q$, *C*^{Ar}-NH), 144.19 (+, CH, CH$_2$CH=CH), 141.47 (C$_q$, *C*^{Ar}), 139.06 (C$_q$, *C*^{Ar}), 138.91 (C$_q$, *C*^{Ar}), 135.94 (C$_q$, *C*^{Ar}), 135.33 (C$_q$, *C*^{Ar}-CH$_2$), 133.49 (+, CH, *C*^{Ar}), 132.72 (+, CH, *C*^{Ar}), 131.05 (+, CH, *C*^{Ar}), 128.73 (+, CH, 2 × *C*^{Ar}), 128.45 (+, CH, 2 × *C*^{Ar}), 128.43 (+, CH, *C*^{Ar}), 127.51 (+, CH, *C*^{Ar}), 125.69 (+, CH, *C*^{Ar}), 124.58 (+, CH, CH$_2$CH=CH), 123.15 (+, CH, *C*^{Ar}), 117.58 (+, CH, *C*^{Ar}), 66.58 (–, CH$_2$, *C*H$_2$Ph), 55.79 (+, CH, *C*HN), 52.55 (+, CH$_3$, O*C*H$_3$), 36.38 (–, CH$_2$, *C*H$_2$CH), 35.38 (–, CH$_2$, *C*^{PC}), 35.35 (–, CH$_2$, *C*^{PC}), 33.12 (–, CH$_2$, *C*^{PC}), 32.93 (–, CH$_2$, *C*^{PC}). – **IR** (ATR): ṽ/cm^{-1} = 3375 (vw), 2923 (w), 2851 (w), 1718 (m), 1655 (w), 1595 (w), 1571 (w), 1511 (w), 1425 (w), 1377 (vw), 1260 (m), 1156 (m), 1082 (w), 1026 (w), 979 (w), 891 (w), 797 (w), 738 (w), 718 (w), 696 (w), 660 (w), 587 (vw), 514 (vw), 454 (w). – **MS** (FAB, 3-NBA), m/z (%): 470 (30) [M + H]$^+$, 376 (11) [M – CO$_2$Me]$^+$, 334 (10) [M – C$_8$H$_7$O$_2$]$^+$, 294 (10) [M – C$_{11}$H$_{11}$O$_2$]$^+$, 274 (40) [M – CO$_2$Me – C$_8$H$_7$O$_2$]$^+$, 132 (100) [C$_{10}$H$_{12}$]$^+$. – **HRMS** (FAB, C$_{30}$H$_{32}$O$_4$N$_1$, [M+H]$^+$): calc. 470.2331, found 470.2330.

1-Benzyl 6-methyl (*E*)-5-((*rac*)-4-amino-13-bromo[2.2]paracyclophane)hex-2-enedioate (179)

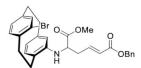

General procedure 4b (GP 4b) was followed by stirring Ru-Complex (**156**, 3.80 mg, 6.00 μmol, 6 mol%) and NaBArF (5.20 mg, 6.00 μmol, 6 mol%) in abs. CH_2Cl_2 (1 mL) in an oven dried vial. 1-Benzyl 6-methyl (*E*)-5-diazohex-2-enedioate (**177e**, 27.4 mg, 0.100 mmol, 1.00 equiv.) and (*rac*)-4-amino-13-bromo[2.2]paracyclophane (**161**, 36.1 mg, 0.120 mmol, 1.20 equiv.) in CH_2Cl_2 (1 mL) was added dropwise, the mixture was stirred at room temperature overnight. The resulting mixture was dried under vacuum and purified directly *via* column chromatography (pentane/Et$_2$O = 3:1) to give the title product **179** as colorless solid, 8.90 mg, 15.0 μmol, 15%.

R_f = 0.25 (pentane/Et$_2$O = 3:1). – **^{1}H NMR** (500 MHz, CDCl$_3$) δ/ppm = 7.48–7.29 (m, 6H, CH$_2$*Ph* + C*H*^{Ar}), 7.08 (ddd, *J* = 15.4, 8.3, 6.8 Hz, 1H, CH$_2$C*H*=CH), 6.71 (d, *J* = 1.6 Hz, 1H, C*H*^{Ar}), 6.47 (dd, *J* = 7.8, 1.6 Hz, 1H, C*H*^{Ar}), 6.45–6.34 (m, 2H, C*H*^{Ar}), 6.14 (dd, *J* = 7.6, 1.6 Hz, 1H, C*H*^{Ar}), 6.05 (d, *J* = 15.6 Hz, 1H, CH$_2$CH=C*H*), 5.42 (d, *J* = 1.5 Hz, 1H, C*H*^{Ar}), 5.22 (s, 2H, C*H*$_2$Ph), 4.11 (t, *J* = 6.4 Hz, 1H, C*H*N), 3.59 (s, 3H, OC*H*$_3$), 3.29 (ddd, *J* = 13.9, 9.3, 2.0 Hz, 1H, C*H*^{PC}), 3.09–2.99 (m, 1H, C*H*^{PC}), 3.00–2.68 (m, 8H, 6 × C*H*^{PC} + C*H*$_2$CH=CH). – **^{13}C NMR** (126 MHz, CDCl$_3$) δ/ppm = 173.53 (C$_q$, *C*O$_2$Me), 165.53 (C$_q$, *C*O$_2$Bn), 145.15 (C$_q$, *C*^{Ar}-NH), 143.44 (+, CH, CH$_2$*C*H=CH), 140.71 (C$_q$, *C*^{Ar}), 140.48 (C$_q$, *C*^{Ar}), 137.72 (C$_q$, *C*^{Ar}), 135.72 (C$_q$, *C*^{Ar}), 135.43 (C$_q$, *C*^{Ar}-CH$_2$), 135.08 (+, CH, *C*^{Ar}), 135.00 (+, CH, *C*^{Ar}), 134.77 (+, CH, *C*^{Ar}), 132.34 (+, CH, *C*^{Ar}), 129.41 (+, CH, *C*^{Ar}), 128.47 (+, CH, 2 × *C*^{Ar}), 128.19 (+, CH, 2 × *C*^{Ar}), 124.30 (+, CH, CH$_2$CH=*C*H), 123.09 (+, CH, *C*^{Ar}), 122.29 (C$_q$, *C*^{Ar}-Br), 115.87 (+, CH, *C*^{Ar}), 66.24 (–, CH$_2$, *C*H$_2$Ph), 55.70 (+, CH, *C*HN), 52.03 (+, CH$_3$, O*C*H$_3$), 35.69 (–, CH$_2$, *C*H$_2$CH), 34.97 (–, CH$_2$, *C*^{PC}), 34.44 (–, CH$_2$, *C*^{PC}), 32.70 (–, CH$_2$, *C*^{PC}), 32.17 (–, CH$_2$, *C*^{PC}). – **IR** (ATR): ṽ/cm^{-1} = 3370 (vw), 2923 (w), 2852 (w), 2117 (vw), 1717 (m), 1656 (w), 1594 (w), 1570 (w), 1511 (w), 1425 (m), 1377 (w), 1262 (m), 1155 (m), 1081 (w), 1030 (w), 977 (m), 904 (vw), 878 (w), 828 (w), 737 (w), 697 (m), 659 (w), 580 (vw), 517 (w), 458 (w). – **MS** (FAB, 3-NBA), m/z (%): 549/547 (38/36) [M + H]$^+$, 500/488 (9/9) [M – CO$_2$Me]$^+$, 414/412 (6/6) [M – C$_8$H$_7$O$_2$]$^+$, 274 (37), 132 (56) [C$_{10}$H$_{12}$]$^+$. – **HRMS** (FAB, C$_{30}$H$_{30}$O$_4$N$_1$^{79}Br$_1$, [M+H]$^+$): calc. 547.1358, found 547.1358.

(*E*)-1-(Benzyloxy)-6-methoxy-1,6-dioxohex-4-en-2-aminium chloride (178fc)

To a solution of **177fc** (67.4 mg, 0.183 mmol, 1.00 equiv.) in MeCN/H$_2$O (1:1, 6.00 mL) was added periodic acid (41.6 mg, 0.183 mmol, 1.00 equiv.) and H$_2$SO$_4$ (1 м, 0.250 mL, 0.250 mmol, 1.40 equiv.). After stirring at room temperature for 3 h, the mixture was washed with CH$_2$Cl$_2$ (3 × 10 mL). The aqueous phase was brought to pH = 10.5 by addition of 5 м aqueous KOH solution and extracted with EtOAc (3 × 10 mL). The combined organic phase was dried over Na$_2$SO$_4$, filtrated, acidified to pH = 1 with HCl (1 м in EtOAc) and concentrated under reduced pressure. The crude was washed with toluene (3 × 2 mL) and dried under vacuum to give the product **178fc** as brown oil, 43.1 mg, 0.140 mmol, 80%.

– **^{1}H NMR** (400 MHz, CD$_3$CN) δ/ppm = 8.63 (brs, 3H, N*H*$_3$), 7.72–7.08 (m, 5H, CH$_2$*Ph*), 6.84 (dt, $^3J_{trans}$ = 15.2, ^4J = 7.4 Hz, 1H, CH$_2$C*H*=), 5.99 (dt, $^3J_{trans}$ = 15.6, ^4J = 1.4 Hz, 1H, =C*H*CO$_2$Bn), 5.39–4.93 (m, 2H, C*H*$_2$Ph), 4.34 (t, ^3J = 6.2 Hz, 1H, C*H*N), 3.65 (s, 3H, OC*H*$_3$), 3.00–2.95 (m, 2H, C*H*$_2$CH=). – **^{13}C NMR** (101 MHz, CD$_3$CN) δ/ppm = 169.58 (C$_q$, *C*O$_2$Me), 167.03 (C$_q$, *C*O$_2$Bn), 141.84 (+, CH, CH$_2$*C*H=), 136.16 (C$_q$, *C*^{Ar}–CH$_2$), 129.73 (+, CH, 2 × *C*^{Ar}), 129.71 (+, CH, *C*^{Ar}), 129.68 (+, CH, 2 × *C*^{Ar}), 126.86 (+, CH, =*C*HCO$_2$Bn), 69.18 (–, CH$_2$, *C*H$_2$Ph), 53.13 (+, CH, *C*HN), 52.38 (+, CH$_3$, O*C*H$_3$), 33.44 (–, CH$_2$, *C*H$_2$CH=). – **IR** (ATR): $\tilde{v}$/cm^{-1} = 2951 (m), 2027 (vw), 1741 (m), 1716 (m), 1657 (w), 1599 (w), 1479 (w), 1442 (m), 1407 (w), 1374 (w), 1315 (w), 1283 (w), 1199 (m), 1155 (m), 1103 (w), 1079 (w), 1030 (w), 987 (m), 909 (w), 855 (w), 749 (w), 732 (m), 696 (m), 574 (w), 546 (vw), 495 (vw), 454 (w). – **MS** (EI, 70 eV), m/z (%):[M – HCl + H]$^+$. – **HRMS** (FAB, C$_{14}$H$_{18}$O$_4$N, [M – HCl + H]$^+$): calc. 264.1236, found 264.1234.

(*E*)-6-(Benzyloxy)-1-methoxy-1,6-dioxohex-4-en-2-aminium chloride (178ec)

To a solution of **177ec** (96.0 mg, 0.260 mmol, 1.00 equiv.) in MeCN/H$_2$O (1:1, 8.00 mL) was added periodic acid (59.3 mg, 0.260 mmol, 1.00 equiv.) and H$_2$SO$_4$ (1 м, 0.360 mL, 0.364 mmol, 1.40 equiv.). After stirring at room temperature for 3 h, the mixture was washed with CH$_2$Cl$_2$ (3 × 10 mL). The aqueous phase was brought to pH = 10.5 by addition of 5 м aqueous KOH solution and extracted with EtOAc (3 × 10 mL). The combined organic phase was dried over Na$_2$SO$_4$, filtrated, acidified to pH = 1 with HCl (1 м in EtOAc) and concentrated under reduced pressure. The crude was washed with toluene (3 × 5 mL) and dried under vacuum to give product **178ec** as brown oil, 62.4 mg, 0.209 mmol, 80%.

– ^{1}H NMR (400 MHz, CD$_3$CN) δ/ppm = 8.10 (s, 3H, NH_3), 7.55–7.16 (m, 5H, CH$_2$Ph), 6.89 (dt, $^3J_{trans}$ = 15.3, ^3J = 7.5 Hz, 1H, CH$_2$CH=), 6 11 (dt, $^3J_{trans}$ = 15.6, ^4J = 1.4 Hz, 1H, =CHCO$_2$Bn), 5.17 (s, 2H, CH_2Ph), 4.25 (t, ^3J = 6.2 Hz, 1H, CHN), 3.77 (s, 3H, OCH_3), 2.95 (td, ^3J = 7.2, ^3J = 5.7 Hz, 2H, CH_2CH=). – ^{13}C NMR (101 MHz, CD$_3$CN) δ/ppm = 169.70 (C$_q$, CO$_2$Me), 166.18 (C$_q$, CO$_2$Bn), 141.94 (+, CH, CH$_2$$C$H=), 137.32 (C$_q$, C^{Ar}), 129.48 (+, CH, 2 × C^{Ar}), 129.09 (+, CH, C^{Ar}), 129.04 (+, CH, 2 × C^{Ar}), 126.90 (+, CH, =CHCO$_2$Bn), 66.91 (–, CH$_2$, CH$_2$Ph), 53.95 (+, CH, CHN), 53.00 (+, CH$_3$, OCH$_3$), 33.04 (–, CH$_2$, CH$_2$CH=). – IR (ATR): $\tilde{v}$/cm^{-1} = 2838 (w), 1741 (w), 1714 (m), 1656 (w), 1586 (vw), 1498 (w), 1438 (w), 1381 (vw), 1358 (vw), 1328 (w), 1280 (w), 1232 (w), 1177 (w), 1152 (w), 1032 (w), 983 (w), 934 (w), 899 (w), 843 (w), 789 (vw), 744 (m), 697 (w), 581 (vw), 559 (vw), 502 (w), 428 (vw), 384 (vw). – MS (FAB, 3-NBA), m/z (%): 264 (100) [M – HCl + H]$^+$. – HRMS (FAB, C$_{14}$H$_{18}$O$_4$N$_1$, [M – HCl + H]$^+$): calc. 264.1236, found 264.1235.

Methyl (E)-5-(((benzyloxy)carbonyl)((E)-buta-1,3-dien-1-yl)amino)pent-2-enoate (179fc)

To a solution of 178fc (30.1 mg, 0.101 mmol, 1.00 equiv.) in 2 mL of abs. CH$_2$Cl$_2$ was added Et$_3$N (15.5 μL, 11.2 mg, 0.111 mmol, 1.10 equiv.). The mixture was stirred at room temperature for 2 h and evaporated under reduced pressure (T = 20 °C). The crude was dissolved in Et$_2$O (2 mL), filtrated and the filtrate was evaporated under reduced pressure (T = 20 °C). The resulting free amine was dissolved in Et$_2$O over 4Å MS, crotonaldehyde (9.21 μL, 7.79 mg, 0.111 mmol, 1.10 equiv.) was added dropwise at 0 °C, the mixture was then brought to room temperature and stirred for 15 h. The crude was filtrated and the filtrate was evaporated under reduced pressure (T = 20 °C) to obtain product 179fc as a brown oil, 28.6 mg, 91.0 μmol, 90%.

– ^{1}H NMR (500 MHz, CDCl$_3$) δ/ppm = 7.82 (d, ^3J = 8.3 Hz, 1H, H^1), 7.35–7.30 (m, 5H, CH$_2$Ph), 6.81 (dt, $^3J_{trans}$ = 15.7, ^3J = 7.3 Hz, 1H, CH^4), 6.39–6.18 (m, 2H, CH^2 + CH^3), 5.85 (dt, $^3J_{trans}$ = 15.6, ^4J = 1.5 Hz, 1H, CH^5), 5.17 (s, 2H, CH_2Ph), 3.93 (dd, ^3J = 7.9, ^3J = 5.5 Hz, 1H, CHN), 3.70 (s, 3H, OCH_3), 2.89–2.66 (m, 2H, CH_2), 1.90 (d, ^3J = 4.9 Hz, 3H, CH_3). – ^{13}C NMR (126 MHz, CDCl$_3$) δ/ppm = 170.93 (C$_q$, C=O), 166.63 (C$_q$, C=O), 166.27 (+, CH, C^1), 144.10 (+, CH, C^5), 143.50 (+, CH, C^3), 135.24 (C$_q$, C^{Ar}), 131.62 (+, CH, C^2), 128.80 (+, CH, C^{Ar}), 128.73 (+, CH, C^{Ar}), 128.70 (+, CH, C^{Ar}), 128.65 (+, CH, C^{Ar}), 128.47 (+, CH, C^{Ar}), 124.08 (+, CH, C^4), 71.62 (+, CH, CHCO$_2$Bn), 67.51 (–, CH$_2$, CH$_2$Ph), 51.60 (+, CH$_3$, OCH$_3$), 36.19 (–, CH$_2$, CH$_2$C^4), 18.67 (+, CH$_3$).

Benzyl (E)-5-(((methoxy)carbonyl)((E)-buta-1,3-dien-1-yl)amino)pent-2-enoate (179ec)

To a solution of **178ec** (128 mg, 0.428 mmol, 1.00 equiv.) in 10 mL of abs. CH_2Cl_2 was added Et_3N (65.6 μL, 11.6 mg, 0.471 mmol, 1.10 equiv.). The mixture was stirred at room temperature for 2 h and evaporated under reduced pressure (T = 20 °C). The resulting crude was dissolved in 15 mL of Et_2O, filtrated, the filtrate was evaporated under reduced pressure (T = 20 °C). The resulting free amine was dissolved in Et_2O over 4Å MS, crotonaldehyde (39.0 μL, 33.0 mg, 0.471 mmol, 1.10 equiv.) was added dropwise at 0 °C, the mixture was then brought to room temperature and stirred for 15 h, filtrated and evaporated under reduced pressure (T = 20 °C) to give product **179ec** as light yellow liquid, 74.4 mg, 0.240 mmol, 55%.

The hydrolysis-sensitive product was used immediately without purification in the next step.

– **^{1}H NMR** (500 MHz, CDCl$_3$) δ/ppm = 7.81 (d, ^3J = 8.1 Hz, 1H, CH^1), 7.47–7.30 (m, 5H, CH$_2$Ph), 6.87 (dt, $^3J_{trans}$ = 15.7, ^3J = 7.3 Hz, 1H, CH^4), 6.35–6.26 (m, 2H, CH^2 + CH^3), 5.92 (dt, $^3J_{trans}$ = 15.7, ^4J = 1.5 Hz, 1H, CH^5), 5.16 (s, 2H, CH_2Ph), 3.89 (dd, ^3J = 8.0, ^3J = 5.5 Hz, 1H, CHN), 3.73 (s, 3H, OCH_3), 2.98–2.58 (m, 2H, CH_2), 1.89 (d, ^3J = 4.6 Hz, 3H, CH_3). – **^{13}C NMR** (126 MHz, CDCl$_3$) δ/ppm = 171.58 (C$_q$, C=O), 166.19 (C$_q$, C=O), 165.99 (+, CH, C^1), 144.56 (+, CH, C^5), 143.50 (+, CH, C^3), 136.10 (C$_q$, C^{Ar}), 131.62 (+, CH, C^2), 128.67 (+, CH, 2 × C^{Ar}), 128.32 (+, CH, C^{Ar}), 128.29 (+, CH, 2 × C^{Ar}), 124.15 (+, CH, C^4), 71.78 (+, CH, CHCO$_2$Me), 66.28 (–, CH$_2$, CH$_2$Ph), 52.58 (+, CH$_3$, OCH$_3$), 36.27 (–, CH$_2$, CH$_2$C^4), 18.68 (+, CH$_3$).

6.2.8 Synthesis and Characterization of Iodo-substituted Ketones

5-Iodopentan-2-one (120a)

According to general procedure 5 (GP 5), sodium iodide (37.5 g, 0.25 mol, 2.50 equiv.) was added to a solution of 5-chloropentan-2-one (12.0 g, 0.10 mol, 1.00 equiv.) in dry acetone (200 mL) and the mixture was stirred under reflux overnight. After removing the solvent udner reduced preesure, the residue was dissolved in water and extracted with CH_2Cl_2 (3 × 100 mL), the combined organic layer was washed with brine, dried over Na_2SO_4, filtrated and concentrated under vacuum. The product **102a** was obtained as dark brown liquid, 19.9 g, 0.940 mmol, 94%. The product was used directly for next step without further purification.

R_f = 0.45 (c-Hex/EtOAc = 6:1). – ^1H NMR (300 MHz, CDCl$_3$) δ/ppm = 3.22 (t, ^3J = 6.6 Hz, 2H, CH_2^1), 2.59 (t, ^3J = 7.0 Hz, 2H, CH$_2$, CH_2^3), 2.17 (s, 3H, CH_3), 2.06 (p, ^3J = 6.8 Hz, 2H, CH_2^2). The analytical data match those reported in the literature.[251]

6-Iodohexan-2-one (120b)

According to general procedure 5 (GP 5), sodium iodide (6.02 g, 40.2 mol, 1.20 equiv.) was added to a solution of 6-chlorohexan-2-one (4.48 g, 33.4 mmol, 1.00 equiv.) in dry acetone (100 mL) and the mixture was stirred under reflux cvernight. After removing the solvent under reduced preesure, the residue was dissolved in water and extracted with CH_2Cl_2 (3 × 50 mL), the combined organic layer was washed with brine, dried over Na_2SO_4, filtrated and concentrated under vacuum. The produc was obtained as yellow liquid, 8.46 g, 0.037 mmol, 66%. The product was used directly for next step without further purification.

R_f = 0.30 (c-Hex/EtOAc = 6:1). – ^1H NMR (300 MHz, CDCl$_3$) δ/ppm = 3.16 (t, ^3J = 6.8 Hz, 2H, CH_2^1), 2.44 (t, ^3J = 7.1 Hz, 1H, CH_2^4), 2.13 (s, 3H, CH_3), 1.83–1.75 (m, 2H, CH_2^2), 1.73–1.63 (m, 1H, CH_2^4). The analytical data match those reported in the literature.[252,253]

4-Iodo-1-phenylbutan-1-one (120c)

According to general procedure 5 (GP 5), sodium iodide (9.37 g, 62.5 mol, 2.50 equiv.) was added to a solution of 4-chlorobutyrophenone (4.55 g, 25.0 mmol, 1.00 equiv.) in dry acetone (100 mL) and the mixture was stirred under reflux overnight. After removing the solvent udner reduced preesure, the residue was dissolved in water and extracted with CH_2Cl_2 (3 × 50 mL), the combined organic layers was washed with brine, dried over Na_2SO_4, filtrated and concentrated under vacuum. The product was obtained as yellow liquid, 6.37 g, 0.930 mmol, 93%. The product was used directly for next step without further purification.

R_f = 0.38 (c-Hex/EtOAc = 6:1). – ^{1}**H NMR** (300 MHz, CDCl$_3$) δ/ppm = 8.19–7.86 (m, 2H, 2 × CH^{Ar}), 7.60–7.53 (m, 1H, CH^{Ar}), 7.50–7.40 (m, 2H, 2 × CH^{Ar}), 3.31 (t, ^3J = 6.6 Hz, 2H, CH_2^1), 3.12 (t, ^3J = 7.0 Hz, 2H, CH_2^3), 2.25 (p, ^3J = 6.8 Hz, 2H, CH_2^2). The analytical data match those reported in the literature.[254]

6.2.9 Synthesis and Characterization of the Hydroxyl-substituted α-Diazocarbonyl Compounds

Methyl 2-acetyl-6-oxoheptanoate (121a)

General procedure 6 (GP 6) was followed by adding methyl acetoacetate (6.48 mL, 6.97 g, 60.0 mmol, 1.20 equiv.) in 30 mL of dry THF to a suspension of sodium hydride (60% in oil, 2.40 g, 60.0 mmol, 1.20 equiv.) in 30 mL of dry THF dropwise. After the mixture turned clear, a solution of 5-iodopentan-2-one (**120a**, 10.6 g, 50.0 mmol, 1.00 equiv.) in 40 mL of dry THF was added dropwise. The solution was allowed to warm to room temperature before heating to reflux overnight. The product **121a** was isolated *via* column chromatography (c-Hex/EtOAc = 3:1) as colorless liquid, 3.25 g, 16.2 mmol, 48%.

R_f = 0.14 (c-Hex/EtOAc = 3:1). – ^{1}**H NMR** (300 MHz, CDCl$_3$) δ/ppm = 3.65 (s, 3H, OCH_3), 3.36 (t, ^3J = 7.3 Hz, 1H, CH), 2.38 (t, ^3J = 7.2 Hz, 2H, CH_2^1), 2.14 (s, 3H, CH_3), 2.04 (s, 3H, CH_3), 1.73 (q, ^3J = 7.5 Hz, 2H, CH_2^3), 1.45 (dt, ^2J = 15.3, ^3J = 7.6 Hz, 2H, CH_2^2). – ^{13}**C NMR** (75 MHz, CDCl$_3$) δ/ppm = 208.00 (C$_q$, CH$_3$C=O), 202.75 (C$_q$, CH$_3$C=O), 169.98 (C$_q$, CO$_2$CH$_3$), 59.31 (+, CH, CHCO), 52.36 (+, CH$_3$, OCH$_3$), 42.98 (–, CH$_2$, C^1), 29.82 (+, CH$_3$, COCH$_3$), 28.88 (+, CH$_3$, COCH$_3$), 27.38 (–, CH$_2$, C^3), 21.32 (–, CH$_2$, C^2). – **IR** (ATR): ṽ/cm^{-1} = 2953 (w), 1739 (m), 1709 (s), 1434 (w), 1358 (m), 1203 (m), 1147 (m), 722 (vw), 597 (vw), 536 (vw), 423 (vw), 384 (vw). – **MS** (EI, 70 eV), m/z (%): 200 (11) [M]$^+$, 157 (17) [M – C$_2$H$_3$O]$^+$, 84 (100) [C$_5$H$_8$O]$^+$. – **HRMS** (EI, C$_{10}$H$_{16}$O$_4$) calc. 200.1049, found 200.1047.

Methyl 2-diazo-6-oxoheptanoate (109a)

General procedure 7 (GP 7) was followed by adding Et$_3$N (6.30 mL, 4.55 g, 45.0 mmol, 3.00 equiv.) to a solution of methyl 2-acetyl-6-oxoheptanoate (**121a**, 3.00 g, 15.0 mmol, 1.00 equiv.) in 40 mL of CH$_3$CN. *p*-ABSA (7.20 g, 30.0 mmol, 2.00 equiv.) in 20 mL of CH$_3$CN was added dropwise at 0 °C. The reaction was stirred at room temperature overnight. The product **109a** was isolated *via* column chromatography (*c*-Hex/EtOAc = 3:1) as yellow liquid, 1.79 g, 9.72 mmol, 60%.

R_f = 0.40 (*c*-Hex/EtOAc = 2:1). – **^{1}H NMR** (400 MHz, CDCl$_3$) δ/ppm = 3.73 (s, 3H, OCH_3),2.48 (t, ^3J = 7.2 Hz, 2H, C$H_2{}^1$), 2.30 (t, ^3J = 7.4 Hz, 2H, C$H_2{}^3$), 2.12 (s, 3H, OCH_3), 1.76 (p, ^3J = 7.3 Hz, 2H, C$H_2{}^2$). – **13C NMR** (101 MHz, CDCl$_3$) δ/ppm = 207.94 (C$_q$, C=O), 52.00 (+, CH$_3$, OCH_3), 42.18 (–, CH$_2$, C^1), 30.05 (+, CH$_3$, COCH$_3$), 22.79 (–, CH$_2$, C^3), 21.85 (–, CH$_2$, C^2). – **IR** (ATR): $\tilde{v}$/cm^{-1} = 2953 (w), 2077 (s), 1684 (s), 1436 (m), 1349 (m), 1306 (m), 1187 (m), 1157 (m), 1116 (s), 1051 (w), 966 (w), 740 (m), 592 (vw), 536 (w), 460 (vw). – **MS** (EI, 70 eV), m/z (%): 156 (81) [M – N$_2$]$^+$, 113 (60) [M – N$_2$ – C$_2$H$_3$O]$^+$, 97 (57) [M – N$_2$ – C$_6$H$_9$O]$^+$. – **HRMS** (EI, C$_8$H$_{12}$O$_3$) calc. 156.0786, found 156.0786. The analytical data matched thosed reported in literature[255].

Methyl 2-diazo-6-hydroxyheptanoate (180ar)

General procedure 8 (GP 8) was followed by adding NaBH$_4$ (56.7 mg, 1.50 mmol, 1.50 equiv.) to a solution of methyl 2-diazo-6-oxoheptanoate (**109a**, 184 mg, 1.00 mmol, 1.00 equiv.) in 10 mL of MeOH in portions at 0 °C. After stirring at 0 °C for 30 min, the solvent was removed by evaporation and the product **180ar** was obtained *via* column chromatography (*c*-Hex/EtOAc = 2:1) as yellow liquid, 180 mg, 0.967 mmol, 97%.

R_f = 0.10 (*c*-Hex/EtOAc = 2:1). – **^{1}H NMR** (500 MHz, CDCl$_3$) δ/ppm = 3.83 (q, ^3J = 6.0 Hz, 1H, CHOH), 3.76 (s, 3H, OCH_3), 2.39–2.27 (m, 2H, C$H_2{}^1$), 1.67–1.57 (m, 2H, C$H_2{}^3$), 1.54–1.39 (m, 2H, C$H_2{}^2$), 1.20 (d, ^3J = 6.2 Hz, 3H, CH_3). – **^{13}C NMR** (126 MHz, CDCl$_3$)) δ/ppm = 168.20 (C$_q$, CO$_2$Me), 67.87 (+, CH, CHOH), 52.05 (+, CH$_3$, OCH_3), 37.99 (–, CH$_2$, C^1), 23.94 (+, CH$_3$, CHCH$_3$), 23.63 (–, CH$_2$, C^3), 23.08 (–, CH$_2$, C^2). – **IR** (ATR): $\tilde{v}$/cm^{-1} = 3410 (w), 2930 (w), 2080 (m), 1684 (m), 1578 (w), 1437 (m), 1347 (m), 1189 (m), 1130 (m), 995 (w), 860 (vw), 820 (vw), 776 (vw), 739 (w), 543 (vw). – **MS** (EI, 70 eV), m/z (%): 158 (1) [M – N$_2$]$^+$, 99 (13) [M – N$_2$ – C$_2$H$_3$O$_2$]$^+$, 59 (60) [C$_2$H$_3$O$_2$]$^+$. – **HRMS** (EI, C$_8$H$_{14}$O$_3$) calc. 158.0943, found 158.0942.

Ethyl 2-acetyl-6-oxoheptanonate (121b)

General procedure 6 (GP 6) was followed by adding ethyl acetoacetate (1.51 mL, 1.59 g, 12.0 mmol, 1.20 equiv.) in 15 mL of dry THF to a suspension of sodium hydride (60% in oil, 0.480 g, 12.0 mmol, 1.20 equiv.) in 10 mL of dry THF dropwise. After the mixture turned clear, a solution of 5-iodopentan-2-one (120a, 2.12 g, 10.0 mmol, 1.00 equiv.) in 20 mL of dry THF was added dropwise. The solution was allowed to warm to room temperature before heating to reflux overnight. The product 121b was isolated *via* column chromatography (*c*-Hex/EtOAc = 3:1) as a colorless liquid, 721 mg, 3.40 mmol, 34%.

R_f = 0.11 (*c*-Hex/EtOAc = 4:1). – **^{1}H NMR** (500 MHz, CDCl$_3$) δ/ppm = 4.18 (q, ^3J = 7.2 Hz, 2H, OC*H*$_2$CH$_3$), 3.40 (t, ^3J = 7.3 Hz, 1H, C*H*), 2.45 (t, ^3J = 7.2 Hz, 2H, C*H*$_2^1$), 2.22 (s, 3H, C*H*$_3$), 2.12 (s, 3H, C*H*$_3$), 1.88–1.66 (m, 2H, C*H*$_2^3$), 1.54 (p, ^3J = 7.3 Hz, 2H, C*H*$_2^2$), 1.26 (t, ^3J = 7.1 Hz, 3H, OCH$_2$C*H*$_3$). – **^{13}C NMR** (126 MHz, CDCl$_3$)) δ/ppm = 208.22 (C$_q$, *C*=O), 203.03 (C$_q$, *C*=O), 169.69 (C$_q$, *C*O$_2$Et), 61.57 (–, CH$_2$, O*C*H$_2$CH$_3$), 59.76 (+, CH, *C*HCO$_2$Et), 43.24 (–, CH$_2$, *C*^{1}), 30.06 (+, CH$_3$, CO*C*H$_3$), 29.04 (+, CH$_3$, CO*C*H$_3$), 27.52 (–, CH$_2$, *C*^{3}), 21.50 (–, CH$_2$, *C*^{2}), 14.21 (+, CH$_3$, OCH$_2$*C*H$_3$).

Ethyl 2-diazo-6-oxoheptanoate (109b)

General procedure 7 (GP 7) was followed by adding Et$_3$N (1.25 mL, 983 mg, 9.72 mmol, 3.00 equiv.) to a solution of methyl 2-acetyl-6-oxoheptanoate (121b, 695 mg, 3.24 mmol, 1.00 equiv.) in 20 mL of CH$_3$CN. *p*-ABSA (1.55 g, 6.48 mmol, 2.00 equiv.) in 10 mL of CH$_3$CN was added dropwise at 0 °C. The reaction was stirred at room temperature overnight. The product 109b was isolated *via* column chromatography (*c*-Hex/EtOAc = 3:1) as yellow liquid, 298 mg, 1.50 mmol, 46%.

R_f = 0.22 (*c*-Hex/EtOAc = 3:1). – **^{1}H NMR** (400 MHz, CDCl$_3$) δ/ppm = 4.19 (q, ^3J = 7.1 Hz, 2H, OC*H*$_2$CH$_3$), 2.48 (t, ^3J = 7.2 Hz, 2H, C*H*$_2^1$), 2.30 (t, J = 7.4 Hz, 2H, C*H*$_2^3$), 2.13 (s, 3H, COC*H*$_3$), 1.82–1.71 (m, 2H, C*H*$_2^2$), 1.25 (t, ^3J = 7.1 Hz, 3H, OCH$_2$C*H*$_3$). – **^{13}C NMR** (101 MHz, CDCl$_3$) δ/ppm = 208.02 (C$_q$, *C*=O), 167.59 (C$_q$, *C*O$_2$Et), 60.90 (–, CH$_2$, O*C*H$_2$CH$_3$), 42.24 (–, CH$_2$, *C*^{1}), 30.09 (+, CH$_3$, CO*C*H$_3$), 22.77 (–, CH$_2$, *C*^{3}), 21.88 (–, CH$_2$, *C*^{4}), 14.61 (+, CH$_3$, OCH$_2$*C*H$_3$). – **IR** (ATR): ṽ/cm^{-1} = 2934 (vw), 2078 (m), 1682 (m), 1395 (w), 1369 (w), 1303 (w), 1156 (m), 1113 (m), 1051 (w), 1020 (w), 740 (w), 591 (vw), 535 (vw), 432 (vw). – **MS** (EI, 70 eV), m/z (%): 232 (87) [M – N$_2$]$^+$, 97 (62) [C$_6$H$_9$O]$^+$, 55 (100) [C$_5$H$_6$]$^+$. – **HRMS** (EI, C$_{19}$H$_{14}$O$_3$) calc. 170.0943, found 170.0944.

Ethyl 2-diazo-6-hydroxyheptanoate (180br)

General procedure 8 (GP 8) was followed by adding NaBH$_4$ (6.00 mg, 0.15 mmol, 1.50 equiv.) to a solution of ethyl 2-diazo-6-oxoheptanoate (**109b**, 19.8 mg, 0.100 mmol, 1.00 equiv.) in 5 mL of MeOH in portions at 0 °C. After stirring at 0 °C for 30 min, the solvent was removed by evaporation and the product was obtained *via* column chromatography (*c*-Hex/EtOAc = 2:1) as yellow liquid, 19.0 mg, 0.950 mmol, 95%

R_f = 0.15   (*c*-Hex/EtOAc = 2:1). – **^{1}H NMR**   (400 MHz,   CDCl$_3$)   δ/ppm = 4.22 (q, ^3J = 7.1 Hz,   2H,   CH_2CH$_3$),   3.83   (q, ^3J = 6.1 Hz,   1H,   CHOH),   2.34 (td, ^3J = 7.6, ^4J = 1.9 Hz, 2H, CH$_2$^{1}), 1.69–1.53 (m, 2H, CH$_2$^{3}), 1.54–1.44 (m, 2H, CH$_2$^{2}), 1.28 (d, ^3J = 7.0 Hz, 3H, CH$_2$CH_3), 1.20 (d, ^3J = 6.3 Hz, 3H, CHCH_3). – **^{13}C NMR** (101 MHz, CDCl$_3$) δ/ppm = 167.10 (C$_q$, CO_2Bn), 68.19 (+, CH, CHOH), 61.23 (–, CH$_2$, CH$_2$CH$_3$), 38.53 (–, CH$_2$, C^1), 30.17 (–, CH$_2$, C^3), 24.14 (+, CH$_3$, CHCH_3), 23.57 (–, CH$_2$, C^2), 14.99 (+, CH$_2$, CH$_2$CH$_3$).

Benzyl 2-acetyl-6-oxoheptanoate (121c)

General procedure 6 (GP 6) was followed by adding benzyl acetoacetate (2.07 mL, 2.31 g, 12.0 mmol, 1.20 equiv.) in 20 mL of dry THF to a suspension of sodium hydride (60% in oil, 0.45 g, 12.0 mmol, 1.20 equiv.) in 10 mL of dry THF dropwise. After the mixture turned clear, a solution of 5-iodopentan-2-one (**120a**, 2.12 g, 10.0 mmol, 1.00 equiv.) in 20 mL of dry THF was added dropwise. The solution was allowed to warm to room temperature before heating to reflux overnight. The product **121c** was isolated *via* column chromatography (*c*-Hex/EtOAc = 3:1) as a colorless liquid, 943 mg, 3.41 mmol, 34%.

R_f = 0.22 (*c*-Hex/EtOAc = 3:1). – **^{1}H NMR** (400 MHz, CDCl$_3$) δ/ppm = 7.38–7.28 (m, 5H, CH$_2$*Ph*), 5.15 (d, ^4J = 2.3 Hz, 2H, CH_2Ph), 3.45 (t, ^3J = 7.3 Hz, 1H, COCH), 2.42 (t, J = 7.2 Hz, 2H, CH$_2$^{1}), 2.16 (s, 3H, CH_3), 2.09 (s, 3H, CH_3,), 1.86–1.78 (m, 2H, CH$_2$^{3}), 1.57–1.46 (m, 2H, CH$_2$^{2}). – **^{13}C NMR** (101 MHz, CDCl$_3$) δ/ppm = 208.05 (C$_q$, C=O), 202.66 (C$_q$, C=O), 169.48 (C$_q$, CO_2Bn), 135.35 (C$_q$, C^{Ar}), 128.72 (+, CH, 2 × C^{Ar}), 128.58 (+, CH, C^{Ar}), 128.44 (+, CH, 2 × C^{Ar}), 67.24 (–, CH$_2$, CH$_2$Ph), 59.66 (+, CH, COCH), 43.14 (–, CH$_2$, C^1), 29.96 02 (+, CH$_3$, COCH$_3$), 29.02 (+, CH$_3$, COCH$_3$), 27.50 (–, CH$_2$, C^3), 21.46 (–, CH$_2$, C^2). – **IR** (ATR): ṽ/cm^{-1} = 2924 (w), 1738 (m), 1710 (s), 1498 (vw), 1455 (w), 1358 (m), 1142 (m), 1081 (w), 963 (w), 750 (w), 698 (m), 597 (w), 500 (vw). – **MS** (EI, 70 eV), m/z (%): 276 (1) [M]$^+$, 142 (20) [C$_8$H$_{14}$O$_2$]$^+$, 91 (100) [C$_7$H$_7$]$^+$. – **HRMS** (EI, C$_{16}$H$_{20}$O$_4$) calc. 276.1362, found 273.1361.

Benzyl 2-diazo-6-oxoheptanoate (109c)

General procedure 7 (GP 7) was followed by adding Et$_3$N (1.24 mL, 977 mg, 9.66 mmol, 3.00 equiv.) to a solution of benzyl 2-acetyl-6-oxoheptanoate (**121c**, 943 mg, 3.22 mmol, 1.00 equiv.) in 20 mL of CH$_3$CN. *p*-ABSA (1.55 g, 6.48 mmol, 2.00 equiv.) in 10 mL of CH$_3$CN was added dropwise at 0 °C. The product **109c** was isolated *via* column chromatography (*c*-Hex/EtOAc = 3:1) as yellow liquid, 493 mg, 1.90 mmol, 59%.

R_f = 0.26 (*c*-Hex/EtOAc = 3:1). – **^{1}H NMR** (500 MHz, CDCl$_3$) δ/ppm = 7.44–7.26 (m, 5H, CH$_2$*Ph*), 5.20 (s, 2H, CH$_2$Ph), 2.49 (t, ^3J = 7.2 Hz, 2H, C*H$_2$*^{1}), 2.34 (t, ^3J = 7.4 Hz, 1H, C*H$_2$*^{3}), 2.12 (s, 3H, C*H$_3$*), 1.79 (p, ^3J = 7.3 Hz, 2H, C*H$_2$*^{2}). – **^{13}C NMR** (126 MHz, CDCl$_3$) δ/ppm = 207.96 (C$_q$, *C*=O), 136.17 (C$_q$, *C*^{Ar}), 128.66 (+, CH, 2 × *C*^{Ar}), 128.32 (+, CH, *C*^{Ar}), 128.15 (+, CH, 2 × *C*^{Ar}), 65.43 (–, CH$_2$, *C*H$_2$Ph), 42.19 (–, CH$_2$, *C*^{1}), 30.08 (+, CH$_3$, CO*C*H$_3$), 22.83 (–, CH$_2$, *C*^{3}), 21.85 (–, CH$_2$, *C*^{2}). – **IR** (ATR): ṽ/cm^{-1} = 2943 (vw), 2080 (m), 1683 (s), 1497 (w), 1454 (w), 1380 (m), 1336 (m), 1301 (m), 1212 (w), 1155 (m), 1109 (m), 1048 (w), 958 (w), 913 (w), 737 (m), 698 (m), 594 (w), 506 (w). – **MS** (EI, 70 eV), m/z (%): 232 (6) [M – N$_2$]$^+$, 91 (100) [C$_7$H$_7$]$^+$. – **HRMS** (EI, C$_{14}$H$_{16}$O$_3$) calc. 232.1099, found 232.1100.

Benzyl 2-diazo-6-hydroxyheptanoate (180cr)

General procedure 8 (GP 8) was followed by adding NaBH$_4$ (3.00 mg, 75.0 *μ*mol, 1.50 equiv.) to a solution of benzyl 2-diazo-6-oxoheptanoate (**109c**, 13.0 mg, 50.0 *μ*mol, 1.00 equiv.) in 10 mL of MeOH in portions at 0 °C. After stirring at 0 °C for 30 min, the solvent was removed by evaporation and the product **180cr** was obtained *via* column chromatography (*c*-Hex/EtOAc = 2:1) as yellow liquid, 11.4 mg, 3.5 *μ*mol, 87%.

R_f = 0.28 (*c*-Hex/EtOAc = 2:1). – **^{1}H NMR** (400 MHz, CDCl$_3$) δ/ppm = 7.42–7.29 (m, 5H, CH$_2$*Ph*), 5.21 (s, 2H, CH$_2$Ph), 3.82 (q, ^3J = 6.1 Hz, 1H, C*H*OH), 2.36 (td, ^3J = 7.6, ^3J = 1.7 Hz, 2H, C*H$_2$*^{1}), 1.62–1.56 (m, 2H, C*H$_2$*^{3}), 1.53–1.46 (m, 2H, C*H$_2$*^{2}), 1.19 (d, ^3J = 6.2 Hz, 3H, CHC*H$_3$*). – **^{13}C NMR** (101 MHz, CDCl$_3$) δ/ppm = 150.40 (C$_q$, *C*O$_2$Bn), 136.28 (C$_q$, *C*^{Ar}), 128.69 (+, CH, 2 × *C*^{Ar}), 128.34 (+, CH, *C*^{Ar}), 128.18 (+, CH, 2 × *C*^{Ar}), 67.85 (+, CH, *C*HOH), 66.49 (–, CH$_2$, *C*H$_2$Ph), 38.19 (–, CH$_2$, *C*^{1}), 24.13 (–, CH$_2$, *C*^{3}), 23.83 (+, CH$_3$, CH*C*H$_3$), 23.31 (–, CH$_2$, *C*^{2}). – **IR** (ATR): ṽ/cm^{-1} = 3433 (vw), 2926 (w), 2082 (m), 1688 (m), 1497 (vw), 1455 (w), 1379 (w), 1262 (w), 1164 (m), 1124 (m), 818 (vw), 736 (m), 696 (m), 595 (w), 506 (vw). – **MS** (EI, 70 eV), m/z (%): 234 (4)

$[M - N_2]^+$, 216 (2%) $[M - N_2-H_2O]^+$, 91 (100) $[C_7H_7]^+$. – **HRMS** (EI, $C_{14}H_8O_3$) calc. 234.1256, found 234.1254.

Benzyl 2-acetyl-7-oxooctanoate (121d)

General procedure 6 (GP 6) was followed by adding benzyl acetoacetate (7.76 mL, 8.63 g, 44.9 mmol, 1.20 equiv.) in 40 mL of dry THF to a suspension of sodium hydride (60% in oil, 1.80 g, 44.9 mmol, 1.20 equiv.) in 20 mL of dry THF dropwise. After the mixture turned clear, a solution of 6-iodohexane-2-one (**102b**, 8.46 g, 37.4 mmol, 1.00 equiv.) in 40 mL of dry THF was added dropwise. The solution was allowed to warm to room temperature before heating to reflux overnight. The product **121d** was isolated *via* column chromatography (*c*-Hex/EtOAc = 3:1) as a colorless liquid, 4.05 g, 14.0 mmol, 37%.

R_f = 0.22 (*c*-Hex/EtOAc = 3:1). – **^{1}H NMR** (500 MHz, CDCl$_3$) δ/ppm = 7.47–7.29 (m, 5H, CH$_2$*Ph*), 5.16 (s, 2H, C*H$_2$*Ph), 3.45 (t, ^3J = 7.3 Hz, 1H, C*H*COCH$_3$), 2.39 (t, ^3J = 7.4 Hz, 2H, CH$_2^1$), 2.17 (s, 3H, COC*H$_3$*), 2.11 (s, 3H, COC*H$_3$*), 1.93–1.69 (m, 2H, CH$_2^4$), 1.56 (pd, ^3J = 7.4, ^4J = 1.7 Hz, 2H, CH$_2^3$), 1.25 (p, ^3J = 7.8 Hz, 2H, CH$_2^2$). – **^{13}C NMR** (126 MHz, CDCl$_3$) δ/ppm = 208.72 (C$_q$, *C*=O), 202.96 (C$_q$, *C*=O), 169.69 (C$_q$, *C*O$_2$Bn), 135.42 (C$_q$, *C*^{Ar}), 128.75 (+, CH, 2 × *C*^{Ar}), 128.61 (+, CH, *C*^{Ar}), 128.48 (+, CH, 2 × *C*^{Ar}), 67.22 (–, CH$_2$, *C*H$_2$Ph), 59.64 (+, CH, *C*HCOCH$_3$), 43.26 (–, CH$_2$, *C*^{1}), 30.06 (+, CH$_3$, CO*C*H$_3$), 29.07 (+, CH$_3$, CO*C*H$_3$), 27.96 (–, CH$_2$, *C*^{4}), 26.90 (–, CH$_2$, *C*^{3}), 23.38 (–, CH$_2$, *C*^{2}). – **IR** (ATR): ṽ/cm^{-1} = 2936 (w), 1738 (m), 1710 (s), 1498 (vw), 1455 (w), 1357 (m), 1139 (m), 958 (w), 741 (w), 698 (m), 594 (w), 493 (vw). – **MS** (EI, 70 eV), m/z (%): 290 (1) [M]$^+$, 91 (100) [C$_7$H$_7$]$^+$. – **HRMS** (EI, $C_{17}H_{22}O_4$) calc. 290.1518, found 290.1519.

Benzyl 2-diazo-7-oxooctanoate (109d)

General procedure 7 (GP 7) was followed by adding Et$_3$N (4.03 mL, 3.19 g, 31.5 mmol, 3.00 equiv.) to a solution of benzyl 2-acetyl-7-oxooctanoate (**121d**, 3.04 g, 10.5 mmol 1.00 equiv.) in 50 mL of CH$_3$CN. *p*-ABSA (4.53 g, 18.9 mmol 2.00 equiv.) in 20 mL of CH$_3$CN was added dropwise at 0 °C. The product **109d** was isolated *via* column chromatography (*c*-Hex/EtOAc = 3:1) as yellow liquid, 1.57 g, 5.70 mmol, 54%.

R_f = 0.21 (*c*-Hex/EtOAc = 4:1). – **^{1}H NMR** (500 MHz, CDCl$_3$) δ/ppm = 7.38–7.27 (m, 5H, CH$_2$*Ph*), 5.19 (s, 2H, C*H$_2$*Ph), 2.44 (t, ^3J = 7.1 Hz, 2H, CH$_2^1$), 2.32 (t, ^3J = 7.3 Hz, 2H, CH$_2^4$), 2.11 (s, 3H, COC*H$_3$*), 1.61 (p, ^3J = 6.9 Hz, 2H, CH$_2^3$), 1.50 (p, ^3J = 7.3 Hz, 2H, CH$_2^2$). – **^{13}C NMR** (101 MHz, CDCl$_3$) δ/ppm = 208.52 (C$_q$, C=O), 208.49 (C$_q$, C=O),

167.35 (C$_q$, CO$_2$Bn), 136.17 (C$_q$, C^{Ar}), 128.63 (+, 2 × C^{Ar}), 128.23 (+, CH, C^{Ar}), 128.06 (+, 2 × C^{Ar}), 66.38 (–, CH$_2$, CH$_2$Ph), 43.09 (–, CH$_2$, C^1), 29.99 (+, CH$_3$, COCH$_3$), 27.16 (–, CH$_2$, C^4), 23.03 (–, CH$_2$, C^3), 22.70 (–, CH$_2$, C^2). – **IR** (ATR): ṽ/cm^{-1} = 2936 (w), 2078 (m), 1683 (s), 1497 (vw), 1455 (w), 1380 (m), 1300 (m), 1211 (w), 1149 (m), 1110 (m), 1058 (w), 912 (w), 737 (m), 698 (m), 594 (w), 506 (w). – **MS** (EI, 70 eV), m/z (%): 246 (1) [M – N$_2$]$^+$, 91 (100) [C$_7$H$_7$]$^+$. – **HRMS** (EI, C$_{15}$H$_{18}$O$_2$) calc 246.1256, found 246.1257.

Benzyl 2-diazo-7-hydroxyoctanoate (180dr)

General procedure 8 (GP 8) was followed by adding NaBH$_4$ (168 mg, 4.44 mmol, 1.50 equiv.) to a solution of benzyl 2-diazo-7-oxooctanoate (**109d**, 811 mg, 2.96 mmol, 1.00 equiv.) in 5 mL of MeOH in portions at 0 °C. After stirring at 0 °C for 30 min, the solvent was removed by evaporation and the product was obtained *via* column chromatography (*c*-Hex/EtOAc = 2:1) as a yellow liquid, 805 mg, 2.92 mmol, 98%.

R_f = 0.24 (*c*-Hex/EtOAc = 2:1). – **^{1}H NMR** (400 MHz, CDCl$_3$) δ/ppm = 8.00–6.70 (m, 5H, CH$_2$*Ph*), 5.20 (s, 2H, C*H$_2$*Ph), 3.78 (q, ^3J = 6.0 Hz, 1H, C*H*OH), 2.34 (t, ^3J = 7.3 Hz, 2H, C*H$_2$*^{4}), 1.58–1.48 (m, 2H, C*H$_2$*^{1}), 1.48–1.33 (m, 4H, C*H$_2$*^{2} + C*H$_2$*^{3}), 1.18 (d, ^3J = 6.2 Hz, 3H, CHC*H$_3$*). – **^{13}C NMR** (101 MHz, CDCl$_3$)) δ/ppm = 167.67 (C$_q$, CO$_2$Bn), 136.28 (C$_q$, C^{Ar}), 128.66 (+, CH, 2C, C^{Ar}), 128.30 (+, CH, C^{Ar}), 128.14 (+, CH, 2 × C^{Ar}), 67.99 (+, CH, CHOH), 66.44 (–, CH$_2$, CH$_2$Ph), 38.92 (–, CH$_2$, C^1), 27.7 (–, CH$_2$, C^4), 24.94 (–, CH$_2$, C^2), 23.70 (+, CH$_3$, CHCH$_3$), 23.20 (–, CH$_2$, C^3). – **IR** (ATR): ṽ/cm^{-1} = 3399 (vw), 3033 (vw), 2929 (w), 2859 (w), 2078 (m), 1684 (m), 1497 (vw), 1455 (w), 1380 (m), 1300 (m), 1211 (w), 1165 (m), 1123 (m), 1074 (m), 1027 (w), 943 (w), 840 (vw), 736 (m), 696 (m), 595 (w), 506 (w). – **MS** (EI, 70 eV), m/z (%): 248 (1) [M – N$_2$]$^+$, 231 (2) [M – N$_2$ – OH]$^+$, 202 (4) [M – N$_2$ – C$_2$H$_6$O]$^+$, 91 (100) [C$_7$H$_7$]$^+$. – **HRMS** (EI, C$_{15}$H$_{20}$O$_3$) calc. 248.1412, found 248.1414.

Benzyl 7-methyloxepane-2-carboxylate (181d)

General procedure 9 (GP 9) was followed by adding rhodium acetate (0.88 mg, 2.00 μmo., 1.00 mol%) to an oven dried vial. Under argon atmosphere, 0.5 mL of abs. toluene was added. Benzyl 2-diazo-7-hydroxyoctanoate ((*rac*)-**180d** or (*R*)-**180d**, 55.2 mg, 0.200 mmol, 1.00 equiv.) in 0.5 mL of abs. toluene was added dropwise. After stirring under reflux or room temperature for 2 h, the solvent was removed under reduced pressue and the resulting mixture was purified *via* column chromatography (*c*-Hex/EtOAc = 4:1) to obtain two different fractions as two diasteromeres.

The HPLC spectra from each reaction and the coresponding data are shown in Figure 38 and Table 19.

R_{f1} = 0.64 (*c*-Hex/EtOAc = 4:1). – ^{1}**H NMR** (400 MHz, CDCl$_3$) δ/ppm = 7.41–7.17 (m, 5H, CH$_2$*Ph*), 5.16 (s, 2H, C*H$_2$*Ph), 4.29 (dd, *J* = 11.7, 5.1 Hz, 1H, C*H*CO$_2$Bn), 3.96 (dqd, ^2J = 10.2, ^3J = 6.3, ^4J = 1.8 Hz, 1H, C*H*CH$_3$), 2.29–2.07 (m, 2H, C*H$_2$*), 1.97–1.64 (m, 4H, 2 × C*H$_2$*), 1.60–1.24 (m, 2H, C*H$_2$*), 1.21 (d, ^3J = 6.3 Hz, 3H, C*H$_3$*). – ^{13}**C NMR** (101 MHz, CDCl$_3$) δ/ppm = 173.82 (C$_q$, *C*O$_2$Bn), 135.98 (C$_q$, *C*^{Ar}), 128.68 (+, CH, 2 × *C*^{Ar}), 128.36 (+, CH, *C*^{Ar}), 128.25 (+, CH, 2 × *C*^{Ar}), 75.12 (+, CH, *C*HCO$_2$Bn), 72.23 (+, CH, *C*HCH$_3$), 66.44 (–, CH$_2$, *C*H$_2$Ph), 38.14 (–, *C*H$_2$), 32.84 (–, *C*H$_2$), 28.44 (–, *C*H$_2$), 25.78 (–, *C*H$_2$), 22.68 (+, CH$_3$, CH*C*H$_3$).

R_{f2} = 0.57 (*c*-Hex/EtOAc = 4:1). – ^{1}**H NMR** (400 MHz, CDCl$_3$) δ/ppm = 7.41–7.17 (m, 5H, CH$_2$*Ph*), 5.14 (s, 2H, C*H$_2$*Ph), 4.17 (dd, *J* = 9.4, 4.3 Hz, 1H, C*H*CO$_2$Bn), 3.73 (dqd, ^2J = 7.9, ^3J = 6.3, ^4J = 3.9 Hz, 1H, C*H*CH$_3$), 2.03 (ddt, ^2J = 14.2, ^3J = 7.2, ^4J = 4.5 Hz, 2H, C*H$_2$*), 1.93–1.81 (m, 4H, 2 × C*H$_2$*), 1.64–1.49 (m, 2H, C*H$_2$*), 1.24 (d, ^3J = 6.3 Hz, 3H, CHC*H$_3$*). – ^{13}**C NMR** (101 MHz, CDCl$_3$) δ/ppm = 172.64 (C$_q$, *C*O$_2$Bn), 135.98 (C$_q$, *C*^{Ar}), 128.66 (+, CH, 2 × *C*^{Ar}), 128.34 (+, CH, *C*^{Ar}), 128.32 (+, CH, 2 × *C*^{Ar}), 78.14 (+, CH, *C*HCO$_2$Bn), 76.95 (+, CH, *C*HCH$_3$), 66.52 (–, CH$_2$, *C*H$_2$Ph), 37.83 (–, *C*H$_2$), 33.55 (–, *C*H$_2$), 25.59 (–, *C*H$_2$), 24.61 (–, *C*H$_2$), 22.92 (+, CH$_3$, CH*C*H$_3$).

– **IR** (ATR): ṽ/cm^{-1} = 2925 (w), 2853 (w), 1743 (m), 1497 (vw), 1454 (w), 1375 (w), 1292 (w), 1250 (w), 1187 (m), 1126 (s), 1064 (w), 1030 (m), 1001 (w), 947 (w), 910 (w), 818 (w), 773 (w), 747 (w), 696 (w), 592 (w), 492 (w), 395 (vw). – **MS** (EI, 70 eV), m/z (%): 248 (2) [M]$^+$, 113 (100) [M – CO$_2$Bn]. – **HRMS** (EI, C$_{15}$H$_{20}$O$_3$) calc 248.1407, found 248.1407.

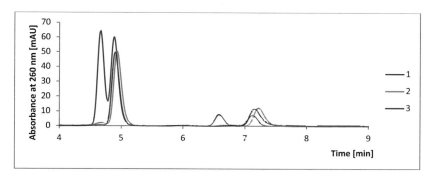

Figure 38. Comparison of the diastereoselectivity and the enantioselectivities of each diastereomer from the O–H insertion product **181d**. 1) reaction with (*rac*)-**180d** as starting material, 80 ºC, 2h. 2) reaction with (*R*)-**180d** as starting material, 80 ºC, 2h. 3) reaction with (*R*)-**180d** as starting material, 80 ºC, r.t., 2h. HPLC condition: PHENOMENEX Amylose 2 column; n-heptane/2-propanol = 98:2; 1.5 mL/min; 30°C; 260 nm UV detector.

Table 19. The original data from the HPLC spectra corresponding to Figure 38.[a]

Entry	Peak	Retention Time [min]	Area [mAU*s]	Percentage [%]
1	1	4.67	497.27	41.71%
	2	4.89	540.52	45.34%
	3	6.58	76.86	6.45%
	4	7.12	77.51	6.50%
2	1	4.68	18.50	2.92%
	2	4.93	459.37	72.59%
	3	6.84	5.65	0.89%
	4	7.22	149.32	23.60%
3	1	4.90	431.42	72.50%
	2	7.15	163.64	27.50%

[a]HPLC condition: PHENOMENEX Amylose 2 column; n-heptane/2-propanol = 98:2; 1.5 mL/min; 30°C; 260 nm UV detector.

Methyl 2-acetyl-6-oxo-6-phenylhexanoate (121e)

General procedure 6 (GP 6) was followed by adding methyl acetoacetate (1.30 mL, 1.39 g, 12.0 mmol, 1.20 equiv.) in 20 mL of dry THF to a suspension of sodium hydride (60% in oil, 0.48 g, 12.0 mmol, 1.20 equiv.) in 10 mL of dry THF dropwise. After the mixture turned clear, a solution of 4-iodo-phenylbutane-2-one (120c, 2.74 g, 10.0 mmol, 1.00 equiv.) in 20 mL of dry THF was added dropwise. The solution was allowed to warm to room temperature before heating to reflux overnight. The product 121e was isolated *via* column chromatography (c-Hex/EtOAc = 4:1) as a colorless liquid, 984 mg, 3.75 mmol, 38%.

R_f = 0.18 (c-Hex/EtOAc = 4:1). – ¹**H NMR** (500 MHz, CDCl₃) δ/ppm = 7.98–7.87 (m, 2H, 2 × CH^{Ar}), 7.60–7.52 (m, 1H, CH^{Ar}), 7.51–7.37 (m, 2H, 2 × CH^{Ar}), 3.73 (s, 3H, OCH₃), 3.50 (t, ^3J = 7.3 Hz, 1H, CHCOCH₃), 3.01 (t, ^3J = 7.1 Hz, 2H, CH₂¹), 2.25 (s, 3H, COCH₃), 2.00–1.87 (m, 2H, CH₂³), 1.81–1.63 (m, 2H, CH₂³). – **¹³C NMR** (101 MHz, CDCl₃)

δ/ppm = 203.02 (C_q, *C*OCH₃), 199.50 (C_q, *C*OPh), 170.22 (C_q, *C*O₂Me), 136.91 (C_q, *C*^Ar),
133.23 (+, CH, *C*^Ar), 128.75 (+, CH, 2 × *C*^Ar), 128.10 (+, CH, 2 × *C*^Ar), 59.66 (+, CH,
*C*HCOCH₃), 52.60 (+, CH₃, O*C*H₃), 38.15 (–, CH₂, *C*¹), 29.11 (+, CH₃, CO*C*H₃), 27.82 (–,
CH₂, *C*³), 21.97 (–, CH₂, *C*²). – **IR** (ATR): ṽ/cm⁻¹ = 2952 (w), 1738 (m), 1712 (m), 1682
(m), 1597 (w), 158 (w), 1447 (w), 1358 (m), 1201 (m), 1146 (m), 1053 (w), 987 (w), 848
(vw), 743 (w), 691 (m), 658 (w), 569 (w). – **MS** (EI, 70 eV), m/z (%): 262 (4) [M]⁺, 146
(33) [C₁₀H₁₀O]⁺, 142 (26) [C₇H₁₀O₃]⁺, 120 (28) [C₈H₈O]⁺, 105 (100) [C₇H₇O]⁺. – **HRMS**
(EI, C₁₅H₁₈O₄) calc. 262.1205, found 262.1206.

Methyl 2-diazo-6-oxo-6-phenylhexanoate (109e)

General procedure 7 (GP 7) was followed by adding Et₃N (1.30 mL,
1.03 g, 10.2 mmol, 3.00 equiv.) to a solution of methyl 2-acetyl-6-
oxo-6-phenylhexanoate (**121e**, 894 mg, 3.40 mmol, 1.00 equiv.) in
20 mL of CH₃CN. *p*-ABSA (1.63 g, 6.80 mmol, 2.00 equiv.) in 20 mL
of CH₃CN was added dropwise at 0 °C. The product **109e** was isolated
via column chromatograph (*c*-Hex/EtOAc = 3:1) as a yellow liquid, 362 mg, 1.47 mmol,
44%.

*R*_f = 0.25 (*c*-Hex/EtOAc = 3:1). – **¹H NMR** (400 MHz, CDCl₃) δ/ppm = 8.08–7.76 (m, 2H,
2 × *C*H^Ar), 7.64–7.49 (m, 1H, *C*H^Ar), 7.51–7.30 (m, 2H, 2 × *C*H^Ar), 3.71 (s, 3H, O*C*H₃),
3.03 (t, ³*J* = 7.1 Hz, 2H, *C*H₂¹), 2.42 (t, ³*J* = 7.4 Hz, 2H, *C*H₂³), 1.97 (p, ³*J* = 7.3 Hz, 2H,
*C*H₂²). – **¹³C NMR** (101 MHz, CDCl₃) δ/ppm = 199.34 (C_q, C=O), 167:94 (C_q, *C*O₂Me),
136.87 (C_q, *C*^Ar), 133.21 (+, CH, *C*^Ar), 128.71 (+, CH, 2 × *C*^Ar), 128.07 (+, CH, 2 × *C*^Ar),
51.99 (+, CH₃, O*C*H₃), 37.17 (–, CH₂, *C*¹), 22.98 (–, CH₂, *C*³), 22.36 (–, CH₂, *C*²).
– **IR** (ATR): ṽ/cm⁻¹ = 2951 (w), 2077 (s), 1680 (s), 1597 (w), 1580 (w), 1436 (m), 1347
(m), 1226 (m), 1190 (m), 1154 (m), 1113 (m), 975 (w), 740 (m), 690 (m), 569 (w). – **MS**
(EI, 70 eV), m/z (%): 218 (37) [M – N₂]⁺, 143 (44) [M – N₂ – C₆H₅]⁺, 105 (100)
[C₆H₅CO]⁺, 77 (55) [C₆H₅]. – **HRMS** (EI, C₁₃H₁₄O₃) calc 246.1099, found 216.1100.

Methyl 2-diazo-6-hydroxy-6-phenylhexanoate (180er)

General procedure 8 (GP 8) was followed by adding NaBH₄ (6.00 mg,
0.150 mmol, 1.50 equiv.) to a solution of methyl 2-diazo-6-oxo-6-
phenylhexanoate (**109e**, 22.0 mg, 0.100 mmol, 1.00 equiv.) in 5 mL
of MeOH in portions at 0 °C. After stirring at 0 °C for 30 min, the
solvent was removed by evaporation and the product was obtained *via* column
chromatography (*c*-Hex/EtOAc = 2:1) as a yellow liquid, 20.5 mg, 82.6 µmol, 83%.

R_f = 0.18 (*c*-Hex/EtOAc = 2:1). – **^{1}H NMR** (400 MHz, CDCl$_3$) δ/ppm = 7.30–7.22 (m, 4H, 4 × CH^{Ar}), 7.19 (ddd, J = 9.3, 3.8, 2.4 Hz, 1H, CH^{Ar}), 4.59 (ddd, J = 7.9, 5.3, 2.6 Hz, 1H, CHOH), 3.65 (s, 3H, OCH_3), 2.24 (t, ^3J = 7.4 Hz, 2H, CH_2^1), 1.81–1.39 (m, 4H, CH_2^3 + CH_2^2). – **^{13}C NMR** (101 MHz, CDCl$_3$) δ/ppm = 168.13 (C$_q$, CO_2Me), 144.65 (C$_q$, C^{Ar}), 128.58 (+, CH, 2 × C^{Ar}), 127.69 (+, CH, C^{Ar}), 125.88 (+, CH, 2 × C^{Ar}), 74.18 (+, CH, CHOH), 51.98 (+, CH$_3$, OCH$_3$), 37.94 (–, CH$_2$, C^1), 24.09 (–, CH$_2$, C^3), 23.07 (–, CH$_2$, C^2). – **IR** (ATR): ṽ/cm^{-1} = 3435 (w), 3028 (vw), 2949 (w), 2862 (w), 2077 (s), 1671 (m), 1493 (w), 1436 (m), 1346 (m), 1306 (m), 1190 (m), 1155 (m), 1118 (m), 1061 (m), 1024 (w), 915 (w), 740 (m), 701 (m), 628 (w), 543 (w). – **MS** (EI, 70 eV), m/z (%): 220 (21) [M – N$_2$]$^+$, 143 (44) [M – N$_2$ – C$_6$H$_5$]$^+$, 107 (76) [C$_7$H$_6$OH]$^+$, 79 (100), 77 (70) [C$_6$H$_5$]. – **HRMS** (EI, C$_{13}$H$_{16}$O$_3$) calc. 220.1099, found 220.1100.

Benzyl 2-acetyl-6-oxo-6-phenylhexanoate (121f)

General procedure 6 (GP 6) was followed by adding a solution of benzyl acetoacetate (2.07 mL, 2.31 g, 12.0 mmol, 1.20 equiv.) in 30 mL of dry THF to a suspension of sodium hydride (60% in oil, 0.48 g, 12.0 mmol, 1.20 equiv.) in 10 mL of dry THF dropwise. After the mixture turned clear, a solution of 4-iodo-phenylbutane-2-one (**120c**, 2.74 g, 10.0 mmol, 1.00 equiv.) in 20 mL of dry THF was added dropwise. The solution was allowed to warm to room temperature before heating to reflux overnight. The product **121f** was isolated *via* column chromatography (*c*-Hex/EtOAc = 5:1) as a colorless liquid, 695 mg, 2.05 mmol, 21%.

R_f = 0.31 (*c*-Hex/EtOAc = 4:1). – **^{1}H NMR** (300 MHz, CDCl$_3$) δ/ppm = 8.07–7.85 (m, 2H, 2 × CH^{Ar}), 7.59–7.51 (m, 1H, CH^{Ar}), 7.45 (ddd, J = 8.4, 6.5 Hz, 1.1 Hz, 2H, 2 × CH^{Ar}), 7.39–7.26 (m, 5H, CH$_2$$Ph$), 5.17 (s, 2H, C$H_2$Ph), 3.48 (t, ^3J = 7.3 Hz, 1H, CHCOCH$_3$), 2.99 (td, ^3J = 7.1, ^4J = 1.1 Hz, 2H, CH_2^1), 2.20 (s, 3H, COCH_3), 2.01–1.89 (m, 2H, CH_2^3), 1.81–1.66 (m, 2H, CH_2^2).

Benzyl 2-diazo-6-oxo-6-phenylhexanoate (109f)

General procedure 7 (GP 7) was followed by adding Et_3N (0.50 mL, 0.389 g, 3.84 mmol, 3.00 equiv.) to a solution of benzyl 2-acetyl-6-oxo-6-phenylhexanoate (121f, 432 mg, 1.28 mmol, 1.00 equiv.) in 10 mL of CH_3CN. *p*-ABSA (0.614 g, 2.56 mmol, 2.00 equiv.) in 50 mL of CH_3CN was added dropwise at 0 °C. The product 109f was isolated *via* column chromatography (*c*-Hex/EtOAc = 6:1 → 4:1) as a yellow liquid, 141 mg, 0.438 mmol, 34%.

R_f = 0.30 (*c*-Hex/EtOAc = 4:1). – **^{1}H NMR** (400 MHz, CDCl$_3$) δ/ppm = 8.00–7.85 (m, 2H, 2 × CH^{Ar}), 7.61–7.52 (m, 1H, CH^{Ar}), 7.46 (dd, *J* = 8.4, 7.0 Hz, 2H, 2 × CH^{Ar}), 7.39–7.27 (m, 5H, CH_2Ph), 5.18 (s, 2H, CH_2Ph), 3.04 (t, ^3J = 7.1 Hz, 2H, CH_2^1), 2.45 (t, ^3J = 7.4 Hz, 2H, CH_2^3), 1.99 (p, ^3J = 7.3 Hz, 2H, CH_2^2). – **^{13}C NMR** (101 MHz, CDCl$_3$) δ/ppm = 199.36 (C$_q$, COPh), 136.90 (C$_q$, C^{Ar}), 136.23 (C$_q$, C^{Ar}), 133.25 (+, CH, C^{Ar}), 128.76 (+, CH, 2 × C^{Ar}), 128.68 (+, CH, 2 × C^{Ar}), 128.32 (+, CH, C^{Ar}), 128.15 (+, CH, 2 × C^{Ar}), 128.13 (+, CH, 2 × C^{Ar}), 66.50 (–, CH$_2$, CH$_2$Ph), 37.22 (–, CH$_2$, C^1), 23.06 (–, CH$_2$, C^3), 22.39 (–, CH$_2$, C^2). – **IR** (ATR): ṽ/cm^{-1} = 2934 (w), 2079 (s), 1671 (vs), 1595 (m), 1580 (m), 1497 (s), 1449 (m), 1415 (s), 1377 (m), 1356 (m), 1316 (m), 1299 (m), 1245 (s), 1202 (m), 1126 (s), 1067 (m), 1029 (m), 974 (m), 727 (s), 683 (s), 655 (m), 591 (m), 572 (m), 523 (s), 480 (w), 454 (m). – **MS** (EI, 70 eV), m/z (%): 294 (20) [M – N$_2$]$^+$, 203 (19), 181 (36), 160 (17), 105 (80) [C$_7$H$_7$O]$^+$, 91 (100) [C$_7$H$_7$]$^+$, 77 (70) [C$_6$H$_5$]$^+$.– **HRMS** (EI, C$_{19}$H$_{18}$O$_3$) calc 294.1256, found 294.1258.

Benzyl 2-diazo-6-hydroxy-6-phenylhexanoate (180fr)

General procedure (GP 8) was followed by adding NaBH$_4$ (57.0 mg, 1.50 mmol, 1.50 equiv.) to a solution of 2-diazo-6-oxo-6-phenylhexanoate (109f, 322 mg, 1.00 mmol, 1.00 equiv.) in 10 mL of MeOH in portions at 0 °C. After stirring for 30 min, the solvent was removed by evaporation and the product was obtained *via* column chromatography (*c*-Hex/EtOAc = 2:1) as yellow liquid, 302 mg, 0.932 mmol, 93%.

R_f = 0.15 (*c*-Hex/EtOAc = 2:1) – **^{1}H NMR** (400 MHz, CDCl$_3$) δ/ppm = 7.41–7.27 (m, 9H, 9 × CH^{Ar}), 5.19 (s, 2H, CH_2Ph), 4.69 (dd, *J* = 7.6, 5.3 Hz, 1H, CHOH), 2.36 (t, ^3J = 7.4 Hz, 2H, CH_2^1), 1.93–1.45 (m, 4H, $CH_2^2 + CH_2^3$). – **^{13}C NMR** (101 MHz, CDCl$_3$) δ/ppm = 167.50 (C$_q$, CO$_2$Bn), 144.62 (C$_q$, C^{Ar}), 136.23 (C$_q$, C^{Ar}), 128.65 (+, CH, 2 × C^{Ar}), 128.29 (+, CH, C^{Ar}), 128.13 (+, CH, 2 × C^{Ar}), 127.78 (+, CH, C^{Ar}), 125.90 (+, CH, 2 × C^{Ar}), 74.26 (+, CH, CHOH), 66.46 (–, CH$_2$, CH$_2$Ph), 37.95 (–, CH$_2$, C^1), 24.14 (–, CH$_2$, C^3), 23.14 (–, CH$_2$, C^2).

7   List of Abbreviations

-	no product
%	percent
(v/v)	volume/volume ratio
3-NBA	3-nitrobenzyl alcohol
abs.	absolute
AcCl	acetyl chloride
AcOH	acetic acid
ADH	alcohol dehydrogenase
API	active pharmaceutical ingredients
ATR	attenuated total reflection
aq.	aqueous
brs	broad singlet
Bn	benzyl
box	bisoxazoline
c	concentration
cap	tetracarprolactamate
Cbz	benzyloxycarbonyl
CBS	COREY-BAKSHI-SHIBATA
CCl_4	tetrachloride carbon
CF_3	trifluromethyl
CH_2Cl_2	dichloromethane
CH_2N_2	diazomethane
CH_3CN	acetonitrile
c-Hex	cyclohexane
Cp	cyclopentadienyl
C_q	quaternary carbon
ClO_4	perchlorate
Cod	1,5-cyclodiene
Cu	copper
CuCl	copper (I) chloride
$Cu(MeCN)_4PF_6$	tetrakis(acetonitrile)copper(I) hexafluorophosphate
Cy	cyclohexyl
d.r.	diastereomeric ratio
DBU	1,8-diazabicyclo[5.4.0]undec-7-ene
DCE	dichloroethane
DEPT	distortionless enhancement by polarization transfer
DFT	density funtional theroy
DMF	N,N-dimethylformamide
DMSO	dimethylsulfoxide
DPEphos	Bis[(2-diphenylphosphino)phenyl] ether

E	electron-withdrawing group
e.g.	exempli gratia (for example)
EDA	ethyl diazoacetate
ee	enantiomeric excess
equiv.	equivalents
Et	ethyl
Et_2O	diethyl ether
Et_3N	triethylamine
EtOAc	ethyl acetate
FAB	fast atom bombardment
Fe	iron
g	gram
GCMS	gas chromatography–mass spectrometry
GDH	glucose dehydrogenase
gem	geminal
GP	general procedure
h	hour
HCl	hydrochloric acid
hept	heptet
HNO_3	nitric acid
HOAc	acetic acid
HPLC	high performance liquid chromatography
HRMS	high resolution mass spectrometry
Hz	hertz
i.e.	id est (that is)
in situ	latin for "on site", without isolation
*i*Pr	isopropyl, pro-2-yl
IR	infrared
J	coupling constant
K_2CO_3	potassium carbonate
KF	potassium fluoride
KRED	ketoreductase
L	liter
LbADH	alcohol dehydrogenase from *Lactobacillus brevis*
LkADH	alcohol dehydrogenase from *kefir*
LiHMDS	lithium bis(trimethylsilyl)amide
m	meta
m	multiplet
m/z	mass-to-charge ratio
MeI	iodomethane
MeOH	methanol
mg	milligram
MHz	mega hertz

min	minute
mL	milliliter
mmol	millimole
MALDI-TOF	Matrix-assisted laser desorption ionization-time of fly
Me	methyl
MOM	methoxylmethyl
MS	mass spectroscopy
MS	molecular sieves
N	normality/ equivalent concentration
n.d.	not determined
n.r.	no reaction
NaBARF	sodium tetrakis[3,5-bis(trifluoromethyl)phenyl]borate
NaBH$_4$	sodium borohydride
NADP$^+$	nicotinamide adenie dinucleotide phosphate (oxidized form)
NADPH	nicotinamide adenine dinucleotide phosphate (reduced form)
NaH	sodium hydride
NaI	sodium iodide
NaOAc	sodium acetate
NaOH	Sodium hydroxide
NBS	N-bromosuccinimide
NH$_4$Cl	ammonium chloride
nBuLi	n-butyllithium
NMR	nuclear magnetic resonance
NOE	nuclear overhauser effect
Nu	nucleophile
o	ortho
^{o}C	degree celsius
OEt	ethoxyl
OMe	methoxyl
p	para
p-ABSA	4-acetamidobenzenesulfonyl azide
PC	[2.2]paracyclophane
PF$_6$	hexafluorophosphate
PG	protecting group
Ph	phenyl
Phen	1,10-phenanthroline
Pheox	phenyloxazoline
pH	potential hydrogen, logarithm of the activity of hydrogen ions
PMP	p-methylphenyl
PPh$_3$	triphenyl phosphine
ppm	parts per million
q	quartet

rac	racemic
$Rh_2(OAc)_4$	rhodium acetate dimer
R_f	retention factor
rpm	rounds per minute
s	seconds
s	singlet
s	strong
sat.	saturated
SE_{Ar}	aromatic electrophile substitution
$SOCl_2$	thionyl chloride
t	triplet
TBAF	tetrabutylammonium fluoride
Tf	triflate
TFA	trifluoroacetic acid
Ts	tosyl
TMS	trimethylsilyl
*t*Bu	*tert*-butyl
THF	tetrahydrofuran
$TiCl_4$	titanium tetrachloride
t_R	retention time
UV	ultraviolet
µg	microgram
µL	microliter
µm	micromole
vs	very strong
vw	very weak
w	weak
xyl	xylene
YGCY1	ketoreductase yGCY1 from yeast
YGre2p	ketoreductase YGre2p from yeast
YGre3p	ketoreductase YGre3p from yeast
yPrpNH6	ketoreductase yPrpNH6 from yeast

8 Literature

[1]   A. K. Yudin, *Catalyzed Carbon-Heteroatom Bond Formation*, Wiley-VCH Verlag & Co. KGaA, Weinheim, Germany, **2010**.

[2]   D. G. Ene, M. P. Doyle, *Chim. Oggi* **1998**, *16*, 37–39. *Recent advances in asymmetric catalytic metal carbene transformations.*

[3]   T. Ye, M. A. McKervey, *Chem. Rev.* **1994**, *94*, 1091–1160. *Organic Synthesis with α-Diazocarbonyl Compounds.*

[4]   S. F. Zhu, Q. L. Zhou, *Acc. Chem. Res.* **2012**, *45*, 1365–1377. *Transition-metal-catalyzed enantioselective heteroatom-hydrogen bond insertion reactions.*

[5]   Z. Zhang, J. Wang, *Tetrahedron* **2008**, *64*, 6577–6605. *Recent studies on the reactions of α-diazocarbonyl compounds.*

[6]   X. Zhao, Y. Zhang, J. Wang, *Chem. Commun.* **2012**, *48*, 10162. *Recent developments in copper-catalyzed reactions of diazo compounds.*

[7]   T. Curtius, *Ber. Dtsch. Chem. Ges.* **1883**, *16*, 2230–2231. *Über die Einwirkung von salpetriger Säure auf salzsauren Glycocolläther.*

[8]   H. Laubmann, *Justus Liebig Ann. Chem.* **1888**, *243*, 244–248. *Über Diazoanhydride: von Ludwig Wolff.*

[9]   F. Arndt, B. Eistert, W. Partale, *Ber. Dtsch. Chem. Ges.* **1927**, *60B*, 1364–1370. *Diazo-methan und o-Nitroverbindungen, II. N-Oxy-isatin aus o-Nitro-benzoylchloride.*

[10]   W. B. Robert Robinson, *J. Chem. Soc.* **1928**, 1310–1318. *CLXXIV.-The Interaction of Benxoyl Chloride and Diaxornethane together with a Discussion of the Reactions.*

[11]   L. D. Proctor, A. J. Warr, *Org. Process Res. Dev.* **2002**, *6*, 884–892. *Development of a continuous process for the industrial generation of diazomethane.*

[12]   K. C. Nicolaou, P. S. Baran, Y. L. Zhong, H. S. Choi, K. C. Fong, Y. He, W. H. Yoon, *Org. Lett.* **1999**, *1*, 883–886. *New synthetic technology for the synthesis of hindered α-diazoketones via acyl mesylates.*

[13]   E. Aller, P. Molina, A. Lorenzo, *Synlett 2000* **2000**, *2000*, 526–528. *N-Isocyanotriphenyliminophosphorane, a Convenient Reagent for the Conversion of Acyl Chlorides into α-Diazoketones.*

[14]   M. M. Bio, G. Javadi, Z. J. Song, *Synthesis (Stuttg).* **2005**, 19–21. *An improved synthesis of N-isocyanoiminotriphenylphosphorane and its use in the preparation of diazoketones.*

[15]   E. Cuevas-Yañez, M. A. García, M. A. De La Mora, J. M. Muchowski, R. Cruz-Almanza, *Tetrahedron Lett.* **2003**, *44*, 4815–4817. *Novel synthesis of α-diazoketones from acyloxyphosphonium salts and diazomethane.*

[16]   M. Regitz, *Angew. Chemie - Int. Ed.* **1967**, *6*, 733–749. *New Methods of Preperative Organic Chemistry Tranfer of Diazo Groups.*

[17]   R. L. Danheiser, R. F. Miller, R. G. Brisbois, S. Z. Park1, *J. Org. Chem.* **1990**, *55*, 1959–1964. *An Improved Method for the Synthesis of a-Diazo Ketones.*

[18]   D. F. Taber, R. B. Sheth, P. V. Joshi, *J. Org. Chem.* **2005**, *70*, 2851–2854. *Simple preparation of α-diazo esters.*

[19]   W. von E. Doering, C. H. DePuy, *J. Am. Chem. Soc.* **1953**, *75*, 5955–5957. *Diazocyclopentadiene.*

[20]   G. G. Hazen, L. M. Weinstock, R. Connell, F. W. Bollinger, *Synth. Commun.* **1981**, *11*, 947–956. *A safer diazotransfer reagent.*

[21]   T. N. Salzmann, R. W. Ratcliffe, B. G. Christensen, F. A. Bouffard, *J. Am. Chem. Soc.* **1980**, *102*, 6161–6163. *A Stereocontrolled Synthesis of (+)-Thienamycin.*

[22]   H. M. L. Davies, H. D. Smith, *Synth. Commun.* **1987**, *17*, 1709–1716. *Diazotransfer reactions with p-acetamidobenzenesulfonyl azide.*

[23]   E. D. Goddard-Borger, R. V. Stick, *Org. Lett.* **2007**, *9*, 3797–3800. *An efficient, inexpensive, and shelf-stable diazotransfer reagent: Imidazole-1-sulfonyl azide hydrochloride.*

[24]   N. Fischer, E. D. Goddard-Borger, R. Greiner, T. M. Klapötke, B. W. Skelton, J. Stierstorfer, *J. Org. Chem.* **2012**, *77*, 1760–1764. *Sensitivities of some imidazole-1-sulfonyl azide salts.*

[25]   C. Peng, J. Cheng, J. Wang, *J. Am. Chem. Soc.* **2007**, *129*, 8708–8709. *Palladium-Catalyzed Cross-Coupling of Aryl or Vinyl Iodides with Ethyl Diazoacetate.*

[26]   F. Ye, S. Qu, L. Zhou, C. Peng, C. Wang, J. Cheng, M. L. Hossain, Y. Liu, Y. Zhang, Z. X. Wang, et al., *J. Am. Chem. Soc.* **2015**, *137*, 4435–4444. *Palladium-Catalyzed C–H Functionalization of Acyldiazomethane and Tandem Cross-Coupling Reactions.*

[27]   M. C. Pirrung, H. Liu, A. T. Morehead, *J. Am. Chem. Soc.* **2002**, *124*, 1014–1023. *Rhodium chemzymes: Michaelis-Menten kinetics in dirhodium(II) carboxylate-catalyzed carbenoid reactions.*

[28]   P. Yates, *J. Am. Chem. Soc.* **1952**, *74*, 5376–5381. *The Copper-catalyzed Decomposition of Diazoketones.*

[29]   C. J. M. David J. Miller, *Tetrahedron* **1995**, *51*, 10811–10843. *Synthetic applications of the O–H insertion reactions of carbenes and carbenoids derived from diazocarbonyl and related diazo compounds.*

[30]   H. Nozaki, H. Takaya, S. Moriuti, R. Noyori, *Tetrahedron* **1968**, *24*, 3655–3669. *Homogeneous catalysis in the decomposition of diazo compounds by copper chelates.*

[31]   R. G. Salomon, J. K. Kochi, *J. Am. Chem. Soc.* **1973**, *95*, 3300–3310. *Copper(I) Catalysis in Cyclopropanations with Diazo Compounds. The Role of Olefin Coordination.*

[32]   R. Paulissen, H. Reimlinger, E. Hayez, A. J. Hubert, P. Teyssié, *Tetrahedron Lett.* **1973**, *14*, 2233–2236. *Transition metal catalysed reactions of diazocompounds - II insertion in the hydroxylic bond.*

[33]   F. Peng, S. J. Danishefsky, *J. Am. Chem. Soc.* **2012**, *134*, 18860–7. *Total synthesis of (±)-maoecrystal V.*

[34]   Y. Liang, H. Zhou, Z.-X. Yu, *J. Am. Chem. Soc.* **2009**, *131*, 17783–17785. *Why Is Copper (I) Complex More Competent Than Dirhodium (II) Complex in Catalytic Asymmetric O–H Insertion Reactions? A Computational Study of the Metal Carbenoid O–H Insertion into Water.*

[35]   J. M. Fraile, J. I. García, V. Martínez-Merino, J. A. Mayoral, L. Salvatella, *J. Am. Chem. Soc.* **2001**, *123*, 7616–7625. *Theoretical (DFT) insights into the mechanism of copper-catalyzed cyclopropanation reactions. Implications for enantioselective catalysis.*

[36]   B. F. Straub, F. Rominger, P. Hofmann, *Organometallics* **2000**, *19*, 4305–4309. *A Fluxional α-Carbonyl Diazoalkane Complex of Copper Relevant to Catalytic Cyclopropanation.*

[37]   I. V. Shishkov, F. Rominger, P. Hofmann, *Organometallics* **2009**, *28*, 1049–1059. *Remarkably stable Copper(I) α-carbonyl carbenes: synthesis, structure, and mechanistic studies of alkene cyclopropanation reactions.*

[38]   M. R. Fructos, T. R. Belderrain, M. C. Nicasio, S. P. Nolan, H. Kaur, M. M. Díaz-Requejo, P. J. Pérez, *J. Am. Chem. Soc.* **2004**, *126*, 10846–10847. *Complete control of the chemoselectivity in catalytic carbene transfer reactions from ethyl diazoacetate: An N-heterocyclic carbene-Cu system that suppresses diazo coupling.*

[39]   D. Gillingham, N. Fei, *Chem. Soc. Rev.* **2013**, *42*, 4918. *Catalytic X–H insertion reactions based on carbenoids.*

[40]   H. Huang, X. Guo, W. Hu, *Angew. Chemie - Int. Ed.* **2007**, *46*, 1337–1339. *Efficient trapping of oxonium ylides with imines: A highly diastereoselective three-component reaction for the synthesis of β-amino-α-hydroxyesters with quaternary stereocenters.*

[41]   Z. Li, B. T. Parr, H. M. L. Davies, *J. Am. Chem. Soc.* **2012**, *134*, 10942–10946. *Highly stereoselective C–C bond formation by rhodium-catalyzed tandem ylide formation/[2,3]-sigmatropic rearrangement between donor/acceptor carbenoids and chiral allylic alcohols.*

[42]   T. Naota, H. Takaya, S. I. Murahashi, *Chem. Rev.* **1998**, *98*, 2599–2660. *Ruthenium-catalyzed reactions for organic synthesis.*

[43]   E. Galardon, P. Le Maux, G. Simonneaux, *J. Chem. Soc. Perkin Trans. 1.* **1997**, 2455–2456. *Insertion of ethyl diazoacetate into N–H and S–H bonds catalyzed by ruthenium porphyrin complexes.*

[44] Q. H. Deng, H. W. Xu, A. W. H. Yuen, Z. J. Xu, C. M. Che, *Org. Lett.* **2008**, *10*, 1529–1532. *Ruthenium-catalyzed one-pot carbenoid N–H insertion reactions and diastereoselective synthesis of prolines.*

[45] M. Austeri, D. Rix, W. Zeghida, J. Lacour, *Org. Lett.* **2011**, *13*, 1394–1397. *CpRu-catalyzed O–H insertion and condensation reactions of α-diazocarbonyl compounds.*

[46] J. Nicoud, H. Boris, *Tetrahedron Lett.* **1971**, 2065–2068. *UNE NOWELLE SYNTHESE ASYMETRIQUE DE L'ALANINE PAR INSERTION D'UN CARBRNE SUR UNE LIAISON N–H.*

[47] C. F. García, M. A. McKervey, T. Ye, *Chem. Commun.* **1996**, 1465–1466. *Asymmetric catalysis of intramolecular N–H insertion reactions of α-diazocarbonyls.*

[48] R. T. Buck, C. J. Moody, A. G. Pepper, *Arkivoc* **2002**, *2002*, 16–33. *N–H Insertion reactions of rhodium carbenoids. Part 4. New chiral dirhodium(II) carboxylate catalysts.*

[49] S. Bachmann, D. Fielenbach, K. A. Jørgensen, *Org. Biomol. Chem.* **2004**, *2*, 3044–3049. *Cu(I)-carbenoid- and Ag(I)-Lewis acid-catalyzed asymmetric intermolecular insertion of a-diazo compounds into N–H bonds.*

[50] B. Liu, S. Zhu, W. Zhang, C. Chen, Q. Zhou, *J. Am. Chem. Soc.* **2007**, 5834–5835. *Highly Enantioselective Insertion of Carbenoids into N–H Bonds Catalyzed by Copper Complexed of Chiral Spiro Bisoxazolines.*

[51] E. C. Lee, G. C. Fu, *J. Am. Chem. Soc.* **2007**, *129*, 12066–12067. *Copper-catalyzed asymmetric N–H insertion reactions: Couplings of diazo compounds with carbamates to generate α-amino acids.*

[52] Z. Hou, J. Wang, P. He, J. Wang, B. Qin, X. Liu, L. Lin, X. Feng, *Angew. Chemie - Int. Ed.* **2010**, *49*, 4763–4766. *Highly enantioselective insertion of carbenoids into N–H bonds catalyzed by copper(I) complexes of binol derivatives.*

[53] B. Xu, S. F. Zhu, X. L. Xie, J. J. Shen, Q. L. Zhou, *Angew. Chemie - Int. Ed.* **2011**, *50*, 11483–11486. *Asymmetric Ni–H insertion reaction cooperatively catalyzed by rhodium and chiral spiro phosphoric acids.*

[54] B. Xu, S. F. Zhu, X. D. Zuo, Z. C. Zhang, Q. L. Zhou, *Angew. Chemie - Int. Ed.* **2014**, *53*, 3913–3916. *Enantioselective N-H insertion reaction of α-aryl α-diazoketones: An efficient route to chiral α-aminoketones.*

[55] E. Aller, G. G. Cox, D. J. Miller, C. J. Moody, *Tetrahedron Lett.* **1994**, *35*, 5949–5952. *Diastereoselectivity in the O–H insertion reactions of rhodium carbenoids derived from phenyldiazoacetates of homochiral alcohols.*

[56] T. C. Maier, G. C. Fu, *J. Am. Chem. Soc.* **2006**, *128*, 4594–4595. *Catalytic enantioselective O–H insertion reactions.*

[57] S. F. Zhu, Y. Cai, H. X. Mao, J. H. Xie, Q. L. Zhou, *Nat. Chem.* **2010**, *2*, 546–551. *Enantioselective iron-catalysed O–H bond insertions.*

[58] S. F. Zhu, B. Xu, G. P. Wang, Q. L. Zhou, *J. Am. Chem. Soc.* **2012**, *134*, 436–442. *Well-defined binuclear chiral spiro copper catalysts for enantioselective N–H insertion.*

[59] S. Kitagaki, K. Sugisaka, C. Mukai, *Org. Biomol. Chem.* **2015**, *13*, 4833–4836. *Synthesis of planar chiral [2.2]paracyclophane-based bisoxazoline ligands bearing no central chirality and application to Cu-catalyzed asymmetric O–H insertion reaction.*

[60] C. J. Brown and A. C. Farthing, *Nature* **1949**, *4178*, 915–916. *Preperation and Structure of Di-p-Xylylene.*

[61] D. J. Cram, H. Steinberg, *J. Am. Chem. Soc.* **1951**, *73*, 5691–5704. *Macro Rings. I. Preparation and Spectra of the Paracyclophanes.*

[62] H. Hopf, *Angew. Chemie - Int. Ed.* **2008**, *47*, 9808–9812. *[2.2]Paracyclophanes in polymer chemistry and materials science.*

[63] S. E. Gibson, J. D. Knight, *Org. Biomol. Chem.* **2003**, *1*, 1256–1269. *[2.2] Paracyclophane derivatives in asymmetric catalysis.*

[64] D. J. Cram, N. L. Allinger, *J. Am. Chem. Soc.* **1955**, *77*, 6289–6294. *Macro Rings. XII. Stereochemical Consequences of Steric Compression in the Smallest Paracyclophane.*

[65] H. Dodziuk, S. Szyma, T. B. Demissie, K. Ruud, P. Ku, H. Hopf, S. Lin, *J. Phys. Chem. A.* **2011**, *115*, 10638–10649. *Structure and NMR Spectra of Some [2.2] Paracyclophanes . The Dilemma of [2.2] Paracyclophane Symmetry.*

[66] K. C. Dewhirst, D. J. Cram, *J. Am. Chem. Soc.* **1958**, *80*, 3115–3125. *Macro Rings. XVII. An Extreme Example of Steric Inhibition of Resonance in a Classically-conjugated Hydrocarbon.*

[67] E. Hedaya, L. M. Kyle, *J. Org. Chem.* **1967**, *32*, 197–199. *Bridge chemistry of paracyclophanes. The mono- and dichloroformylation of [2.2]paracyclophane (di-p-xylylene) with oxalyl chloride.*

[68] D. J. Cram, R. E. Singler, R. C. Helgeson, *J. Am. Chem. Soc.* **1970**, *92*, 7625–7627. *Solvolyses with retention of configuration and cis polar additions in the side-chain chemistry of [2.2]paracyclophane.*

[69] G. Wang, C. Chen, T. Du, W. Zhong, *Adv. Synth. Catal.* **2014**, *356*, 1747–1752. *Metal-free catalytic hydrogenation of imines with recyclable [2.2]paracyclophane-derived frustrated lewis pairs catalysts.*

[70] M. Enders, C. J. Friedmann, P. N. Plessow, A. Bihlmeier, M. Nieger, W. Klopper, S. Bräse, *Chem. Commun.* **2015**, *51*, 5–7. *Unprecedented pseudo-ortho and ortho metallation of [2.2]paracyclophanes – a methyl group matters.*

[71] H. J. Reich, D. J. Cram, *J. Am. Chem. Soc.* **1969**, *91*, 3534–3543. *Macro Rings. XXXVIII. Determination of Positions of Substituents in the [2.2]Paracyclophane Nucleus through Nuclear Magnetic Resonance Spectra.*

[72]   N. Wada, Y. Morisaki, Y. Chujo, *Macromolecules* **2009**, *42*, 1439–1442. *Polymethylenes containing [2.2]paracyclophane in the side chain.*

[73]   G. J. Rowlands, R. J. Seacome, *Beilstein J. Org. Chem.* **2009**, *5*, 1–7. *Enantiospecific synthesis of [2.2]paracyclophane-4-thiol and derivatives.*

[74]   L. Ernst, L. Wittkowski, *European J. Org. Chem.* **1999** *Diastereomers Composed of Two Planar-Chiral Subunits : Bis ([2.2] paracyclophan-4-yl) methane and Analogues.*

[75]   V. V. Kane, A. Gerdes, W. Grahn, L. Ernst, I. Dix, P. G. Jones, H. Hopf, *Tetrahedron Lett.* **2001**, *42*, 373–376. *A novel entry into a new class of cyclophane derivatives: Synthesis of (±)-[2.2]paracyclophane-4-thiol.*

[76]   B. Wang, J. W. Graskemper, L. Qin, S. G. DiMagno, *Angew. Chemie - Int. Ed.* **2010**, *49*, 4079–4083. *Regiospecific reductive elimination from diaryliodonium salts.*

[77]   J. P. Jakob F. Schneider, Roland Frölich, *Synthesis (Stuttg).* **2010**, *20*, 3486–3492. *Synthesis of Enantiopure Planar-Chiral Thiourea Derivatives.*

[78]   E. A. Truesdale, D. J. Cram, *J. Org. Chem.* **1980**, *45*, 3974–3981. *Macro Rings. 49. Use of Transannular Reactions to Add Bridges to [2.2]Paracyclophane.*

[79]   J. E. Gready, T. W. Hambley, K. Kakiuchi, K. Kobiro, Y. Tobe, S. Sternhell, C. W. Tansey, *J. Am. Chem. Soc.* **1990**, *112*, 7537–7540. *NMR Studies of Bond Order in Distorted Aromatic Systems.*

[80]   S. El-Tamany, F. -W. Raulfs, H. Hopf, *Angew. Chemie Int. Ed.* **1983**, *22*, 633–634. *New Link between Cyclophane and Ferrocene Chemistry.*

[81]   M. Brink, *Synthesis (Stuttg).* **1975**, 807–808. *Eine verbesserte Syntheses von [2.2]paracyclophane and 4-formyl[2.2]paracyclophane.*

[82]   H. J. Reich, D. J. Cram, *J. Am. Chem. Soc.* **1968**, *90*, 1365–1367. *Transannular Directive Influences in Electrophilic Substitution of [2.2]Paracyclophane.*

[83]   N. V. Vorontsova, V. I. Rozenberg, E. V. Sergeeva, E. V. Vorontsov, Z. A. Starikova, K. A. Lyssenko, H. Hopf, *Chem. - A Eur. J.* **2008**, *14*, 4600–4617. *Symmetrically tetrasubstituted [2.2]paracyclophanes: Their systematization and regioselective synthesis of several types of bis-bifunctional derivatives by double electrophilic substitution.*

[84]   D. C. Braddock, S. M. Ahmad, G. T. Douglas, *Tetrahedron Lett.* **2004**, *45*, 6583–6585. *A preparative microwave method for the isomerisation of 4,16-dibromo[2.2]paracyclophane into 4,12-dibromo[2.2]paracyclophane.*

[85]   D. J. Cram, D. I. Wilkinson, *J. Am. Chem. Soc.* **1960**, *82*, 5721–5723. *Macro Rings. XXIII. Carbonylchromium Complexes of Paracyclophanes and Model Compounds.*

[86]   E. Langer, H. Lehner, *Tetrahedron* **1973**, *29*, 375–383. *Zur frage transanularer II-II-wechselwirkungen im [2.2]metacyclophan, [2.2]Paracyclophan und 2,2'-spirobiindan.*

[87]   H. H. Aboul Fetouh Mourad, *Tetrahedron Lett.* **1979**, 1209–1212. *Darstellung Neure Chromtricabonyl-komplexe von[2.2]paracyclophane.*

[88]   R. M. and U. Z. Christop Elschenbroich, *Angew. Chemie Int. Ed. English* **1978**, *17*, 531–532. *[2.2]Paracyclophane)chromium(0).*

[89]   E. D. Laganis, R. G. Finke, V. Boekelheide, *Tetrahedron Lett.* **1980**, *21*, 4405–4408. *Multilayered transition metal complexes of cyclophanes.*

[90]   R. Thomas Swan and V. Boekelheide, *Tetrahedron Lett.* **1984**, *25*, 899–900. *A general synthesis of bis(η^6-[2n]cyclophane)ruthenium(II) compounds.*

[91]   and V. B. R. Thomas Swann, A. W. Hanson, *J. Am. Chem. Soc.* **1986**, 3324–3334. *Ruthenium complexes of [2n]cyclophanes. A general synthesis of bis(η^6-[2n]cyclophane)ruthenium(II) compounds and related chemistry.*

[92]   E. D. Laganis, R. H. Voegeli, T. Swann, R. G. Finke, H. Hopf, V. Boekelheide, *Organometallics* **1982**, *31*, 175–184. *A Study of the Synthesis and Properties of Ruthenium Complexes of [2n]cyclophanes.*

[93]   E. D. Laganis, R. G. Finke, V. Boekelheide, R. Thomas Swan and V. Boekelheide, and V. B. R. Thomas Swann, A. W. Hanson, E. D. Laganis, R. H. Voegeli, T. Swann, R. G. Finke, H. Hopf, et al., *Tetrahedron Lett.* **1986**, *25*, 3324–3334. *Ruthenium complexes of [2n]cyclophanes. A general synthesis of bis(η^6-[2n]cyclophane)ruthenium(II) compounds and related chemistry.*

[94]   R. T. Swann, V. Boekelheide, A. W. Hanson, *J. Am. Chem. Soc.* **1984**, *106*, 818–819. *Conversion of η^4, η^6-Bis(arene)ruthenium(0) Complexes to Cyclohexadienyl Analogues of Ruthenocene.*

[95]   P. J. Pye, K. Rossen, R. A. Reamer, N. N. Tsou, R. P. Volante, P. J. Reider, *J. Am. Chem. Soc.* **1997**, *119*, 6207–6208. *A new planar chiral bisphosphine ligand for asymmetric catalysis: Highly enantioselective hydrogenations under mild conditions.*

[96]   P. J. Pye, K. Rossen, R. A. Reamer, R. P. Volante, P. J. Reider, *Tetrahedron Lett.* **1998**, *39*, 4441–4444. *[2.2]PHANEPHOS-ruthenium(II) complexes: Highly active asymmetric catalysts for the hydrogenation of β-ketoesters.*

[97]   N. B. Johnson, I. C. Lennon, P. H. Moran, J. A. Ramsden, *Acc. Chem. Res.* **2007**, *40*, 1291–1299. *Industrial-scale synthesis and applications of asymmetric hydrogenation catalysts.*

[98]   T. M. Konrad, J. A. Fuentes, A. M. Z. Slawin, M. L. Clarke, *Angew. Chemie - Int. Ed.* **2010**, *49*, 9197–9200. *Highly enantioselective hydroxycarbonylation and alkoxycarbonylation of alkenes using dipalladium complexes as precatalysts.*

[99]   C. A. Mullen, A. N. Campbell, Gagne, *Angew. Chemie - Int. Ed.* **2008**, *47*, 6011–6014. *Asymmetric oxidative cation/olefin cyclization of polyenes: Evidence for reversible cascade cyclization.*

[100]   A. P. and A. R. Marchand, A. Maxwell, B. Mootoo, *Tetrahedron* **2000**, *56*, 7331–7228. *Oxazoline mediated routes to a unique amino-acid, 4-Amino-13-carboxy[2.2]paracyclophane, of Planar Chirality.*

[101]   X. W. Wu, K. Yuan, W. Sun, M. J. Zhang, X. L. Hou, *Tetrahedron Asymmetry* **2003**, *14*, 107–112. *Novel planar chiral P,N-[2.2]paracyclophane ligands: Synthesis and application in palladium-catalyzed allylic alkylation.*

[102]   X. Wu, X. Hou, L. Dai, J. Tao, B. Cao, J. Sun, *Tetrahedron:Asymmetry* **2001**, *12*, 529–532. *Synthesis of novel N,O-planar chiral [2,2]paracyclophane ligands and their application as catalysts in the addition of diethylzinc to aldehydes.*

[103]   G. Desimoni, G. Faita, K. A. Jørgensen, *Chem. Rev.* **2011**, *111*, PR284-PR437. *C₂ -Symmetric Chiral Bis(oxazoline) Ligands in Asymmetric Catalysis.*

[104]   T. D. Nelson, A. I. Meyers, *J. Org. Chem.* **1994**, *59*, 2655–2658. *The Asymmetric Ullmann Reaction. 2. The Synthesis of Enantiomerically Pure C₂-Symmetric Binaphthyls.*

[105]   M. B. Andrus, D. Asgari, J. A. Sclafani, *J. Org. Chem.* **1997**, *62*, 9365–9368. *Efficient Synthesis of 1,1'-Binaphthyl and 2,2'-Bi-o-tolyl-2,2'-bis(oxazoline)s and Preliminary Use for the Catalytic Asymmetric Allylic Oxidation of Cyclohexene.*

[106]   Z. Han, Z. Wang, X. Zhang, K. Ding, *Chinese Sci. Bull.* **2010**, *55*, 2840–2846. *Synthesis of novel chiral bisoxazoline ligands with a spiro[4,4]-1,6-nonadiene skeleton.*

[107]   J. Li, G. Chen, Z. Wang, R. Zhang, X. Zhang, K. Ding, *Chem. Sci.* **2011**, *2*, 1141–1144. *Spiro-2,2'-bichroman-based bisoxazoline (SPANbox) ligands for ZnII-catalyzed enantioselective hydroxylation of β-keto esters and 1,3-diester.*

[108]   S.-G. Kim, C.-W. Cho, K. H. Ahn, *Tetrahedron: Asymmetry* **1997**, *8*, 1023–1026. *Synthesis of chiral bis(oxazolinyl)biferrocene ligands and their application to Cu(I)-catalyzed asymmetric cyclopropanation.*

[109]   S.-G. Kim, C.-W. Cho, K. H. Ahn, *Tetrahedron* **1999**, *55*, 10079–10086. *Chiral biferrocene-based bis(oxazolines): Ligands for Cu(I)-catalyzed asymmetric cyclopropanations of ene-diazoacetates.*

[110]   C. Chen, S. F. Zhu, B. Liu, L. X. Wang, Q. L. Zhou, *J. Am. Chem. Soc.* **2007**, *129*, 12616–12617. *Highly enantioselective insertion of carbenoids into O–H bonds of phenols: An efficient approach to chiral α-aryloxycarboxylic esters.*

[111]   B. Liu, S. Zhu, W. Zhang, C. Chen, Q. Zhou, *J. Am. Chem. Soc.* **2007**, 5834–5835. *Highly Enantioselective Insertion of Carbenoids into O–H Bonds.*

[112]   D. S. M. T.Glatzhofer, *J. Mol. Cata A-Chem.* **2000**, *161*, 65–68. *Catalytic enantioselective cyclopropanation of styrene derivatives using N-(2',4'-di-tert-butyl)salicylidene-4-amino[2.2]paracyclophane as an asymmetric ligand.*

[113]   L. X. D. and X. L. H. T. Z. Zhang, *Tetrahedron: Asymmetry* **2007**, *18*, 251–259. *Synthesis of planar chiral [2.2]paracyclophane monophosphine ligands and their application in the umpolung allylation of aldehydes.*

[114]   C. K. Prier, F. H. Arnold, *J. Am. Chem. Soc.* **2015**, *137*, 13992–14006. *Chemomimetic Biocatalysis: Exploiting the Synthetic Potential of Cofactor-Dependent Enzymes To Create New Catalysts.*

[115]   O. M. K. Drauz, H. Gröger, *Enzyme Catalysis in Organic Catalysis*, Wiley-VCH Verlag & Co. KGaA, Weinheim, Germany, **2012**.

[116]   N. J. Turner, L. Humphreys, The Royal Society Of Chemistry, United Kingdom, **2018**, pp. 73–107.

[117]   E. Fischer, *Berichte der Dtsch. Chem. Gesellschaft* **1894**, *27*, 2985–2993. *Einfluss der Configuration auf die Wirkung der Enzyme*.

[118]   P. T. Anastas, J. C. Warner, *Green Chemistry : Theory and Practice*, Oxford University Press, Oxford [England]; New York, **1998**.

[119]   G. P. Moss, https://www.qmul.ac.uk/sbcs/iubmb/, zugegriffen am 17.11.2018.

[120]   *Chirality* **1992**, *4*, 338–340. *FDA'S policy statement for the development of new stereoisomeric drugs*.

[121]   E. J. Corey, R. K. Bakshi, S. Shibata, *J. Am. Chem. Soc.* **1987**, *109*, 5551–5553. *Highly enantioselective borane reduction of ketones catalyzed by chiral oxazaborolidines. Mechanism and synthetic implications*.

[122]   E. J. Corey, S. Shibata, R. K. Bakshi, *J. Org. Chem.* **1988**, *53*, 2861–2863. *An efficient and catalytically enantioselective route to (S)-(-)-phenyloxirane*.

[123]   L. Deloux, M. Srebnik, *Chem. Rev.* **1993**, *93*, 763–784. *Asymmetric boron-catalyzed reactions*.

[124]   J. Pritchard, G. A. Filonenko, R. van Putten, E. J. M. Hensen, E. A. Pidko, *Chem. Soc. Rev.* **2015**, *44*, 3808–3833. *Heterogeneous and homogeneous catalysis for the hydrogenation of carboxylic acid derivatives: history, advances and future directions*.

[125]   M. M. Musa, R. S. Phillips, *Catal. Sci. Technol.* **2011**, *1*, 1311–1323. *Recent advances in alcohol dehydrogenase-catalyzed asymmetric production of hydrophobic alcohols*.

[126]   K. Nakamura, R. Yamanaka, T. Matsuda, T. Harada, *Tetrahedron Asymmetry* **2003**, *14*, 2659–2681. *Recent developments in asymmetric reduction of ketones with biocatalysts*.

[127]   Y. Wang, Z. Chen, A. Mi, W. Hu, *Chem. Commun..* **2004**, 2486–2487. *Novel C–C bond formation through addition of ammonium ylides to arylaldehydes: A facile approach to β-aryl-β-hydroxy α-amino acid frameworks*.

[128]   T. Osako, D. Panichakul, Y. Uozumi, *Org. Lett.* **2012**, *14*, 194–197. *Enantioselective carbenoid insertion into phenolic O–H bonds with a chiral copper(I) imidazoindolephosphine complex*.

[129]   T. M. Beck, B. Breit, *European J. Org. Chem.* **2016**, *2016*, 5839–5844. *Regioselective Rhodium-Catalyzed Addition of β-Keto Esters, β-Keto Amides, and 1,3-Diketones to Internal Alkynes*.

[130]   H. Finkelstein, *Ber. Dtsch. Chem. Ges.* **1910**, *2*, 1528–1532. *Darstellung organischer Jodide aus den entsprechenden Bromiden udn Chloriden*.

182 | Literature

[131] R. Huang, X. Zhou, T. Xu, X. Yang, Y. Liu, *Chem. Biodivers.* **2010**, *7*, 2809–2829. *Diketopiperazines from marine organisms.*

[132] G. Ding, L. Jiang, L. Guo, X. Chen, H. Zhang, Y. Che, *J. Nat. Prod.* **2008**, *71*, 1861–1865. *Pestalazines and Pestalamides , Bioactive Metabolites from the Plant Pathogenic Fungus Pestalotiopsis theae Pestalazines and Pestalamides , Bioactive Metabolites from the Plant Pathogenic Fungus Pestalotiopsis theae.*

[133] D. M. Gardiner, P. Waring, B. J. Howlett, *Microbiology* **2005**, *151*, 1021–1032. *The epipolythiodioxopiperazine (ETP) class of fungal toxins: Distribution, mode of action, functions and biosynthesis.*

[134] U. Gross, M. Nieger, S. Bräse, *Chem. - A Eur. J.* **2010**, *16*, 11624–11631. *A unified strategy targeting the thiodiketopiperazine mycotoxins exserohilone, gliotoxin, the epicoccins, the epicorazines, rostratin a and aranotin.*

[135] S. Zhong, P. F. Sauter, M. Nieger, S. Bräse, *Chem. - A Eur. J.* **2015**, *21*, 11219–11225. *Stereoselective Synthesis of Highly Functionalized Hydroindoles as Building Blocks for Rostratins B-D and Synthesis of the Pentacyclic Core of Rostratin C.*

[136] B. M. Ruff, S. Zhong, M. Nieger, M. Sickert, C. Schneider, S. Bräse, *European J. Org. Chem.* **2011**, 6558–6566. *A combined vinylogous mannich/diels-alder approach for the stereoselective synthesis of highly functionalized hexahydroindoles.*

[137] K. C. Nicolaou, S. Totokotsopoulos, D. Giguère, Y. P. Sun, D. Sarlah, *J. Am. Chem. Soc.* **2011**, *133*, 8150–8153. *Total synthesis of epicoccin G.*

[138] A. Friedrich, M. Jainta, C. F. Nising, S. Bräse, *Synlett* **2008**, *4*, 589–591. *Synthesis of hexahydroindole carboxylic acids by intramolecular Diels-Alder reaction.*

[139] S. Zhong, *Stereoselecktiv Synthese von Hydroindole, Bisindolen Und Thiodiketopiperazinen Als Bausteine Für Rostratin C Sowie Als Neuartige Leitstrukturen Für Die Wirkstoffentwicklung*, Karlsruhe Institut für Technologie (KIT), **2014**.

[140] W. F. Wolfgang Oppolzer, *Helv. Chimica Acta* **1975**, *58*, 590–593. *A Stereoselective Approach to cis- and trans-1,2,3,4,4a,.5,6,8a-Octahydroquinolines by Intramolecular Diels- Alder Reactions.*

[141] A. Padwa, P. E. Yeske, *J. Org. Chem.* **1991**, *56*, 6386–6390. *[3 + 2] Cyclization-Elimination Route to Cyclopentenyl Sulfones Using (Phenylsulfonyl)-1,2-propadiene.*

[142] X. L. Chunming Zhang, *Synlett* **1995**, 645–646. *Umpolung Addition Reaction of Nucleophiles to 2,3-butadienoates Catalyzed by a Phosphine.*

[143] B. M. Trost, C. J. Li, *J. Am. Chem. Soc.* **1994**, *116*, 3167–3168. *Novel "Umpolung" in C-C Bond Formation Catalyzed by Triphenylphosphine.*

[144] T. J. Martin, V. G. Vakhshori, Y. S. Tran, O. Kwon, *Org. Lett.* **2011**, *13*, 2586–2589. *Phosphine-catalyzed β'-umpolung addition of nucleophiles to activated α-alkyl allenes.*

[145]   J.-E. B. Jonas Nyhlen, LARS Eriksson, *Chirality* **2008**, *20*, 47–50. *Synthesis and Optical Resolution of an Allenoic Acid by Diastereomeric Salt Formation Induced by Chiral Alkaloids*.

[146]   E. Kohl-mines, *Helv. Chmica Acta* **1985**, *68*, 2249–2253. *Synthesis of Alkenyl-Substituted Allenecarboxylates*.

[147]   G. Buono, *Tetrahedron Lett.* **1972**, 3257–3259. *Nouvelle voie D'acces aux cetones, esteres, nitriles a-alleniques*.

[148]   M. M. D. Kui Lu, Ohyun Kwon, Kay M. Brummond, *Org. Synth.* **2009**, *86*, 212–224. *Phosphine-catalyzed [4+2]annulation: synthesis of ethyl 6-phenyl-1-tosyl-1,2,5,6-tetrahydropyridine-3-carboxylate*.

[149]   R. Na, C. Jing, Q. Xu, H. Jiang, X. Wu, J. Shi, J. Zhong, M. Wang, D. Benitez, E. Tkatchouk, et al., *J. Am. Chem. Soc.* **2011**, *133*, 13337–13348. *Phosphine-catalyzed annulations of azomethine imines: Allene-dependent [3+2], [3+3], [4+3], and [3+2+3] pathways*.

[150]   D. Klee, N. Weiss, J. Lahann, *Mod. Cyclophane Chem.* **2005**, doi:10.1002/3527603964.ch18. *Vapor-Based Polymerization of Functionalized [2.2]Para-cyclophanes: A Unique Approach towards Surface-Engineered Microenvironments*.

[151]   A. K. Ghosh, P. Mathivanan, J. Cappiello, *Tetrahedron Asymmetry* **1998**, *9*, 1–45. *C2-symmetric chiral bis(oxazoline)-metal complexes in catalytic asymmetric synthesis*.

[152]   J. S. Johnson, D. A. Evans, *Acc. Chem. Res.* **2000**, *33*, 325–335. *Chiral bis(oxazoline) copper(II) complexes: Versatile catalysts for enantioselective cycloaddition, aldol, Michael, and carbonyl ene reactions*.

[153]   R. Rasappan, D. Laventine, O. Reiser, *Coord. Chem. Rev.* **2008**, *252*, 702–714. *Metal-bis(oxazoline) complexes: From coordination chemistry to asymmetric catalysis*.

[154]   H. Nishiyama, H. Sakaguchi, T. Nakamura, M. Horihata, M. Kondo, K. Itoh, *Organometallics* **1989**, *8*, 846–848. *Chiral and C2-symmetrical bis(oxazolinyl-pyridine)rhodium(III) complexes: effective catalysts for asymmetric hydrosilylation of ketones*.

[155]   E. J. Corey, N. Imai, H. Y. Zhang, *J. Am. Chem. Soc.* **1991**, *113*, 728–729. *Designed Catalyst for Enantioselective Diels-Alder Addition from a C2-Symmetric Chiral Bis(oxazoline)-Fe(III) Complex*.

[156]   D. A. Evans, K. A. Woerpel, M. M. Hinman, M. M. Faul, *J. Am. Chem. Soc.* **1991**, *113*, 726–728. *Bis(oxazolines) as Chiral Ligands in Metal-Catalyzed Asymmetric Reactions. Catalytic, Asymmetric Cyclopropanation of Olefins*.

[157]   E. V Sergeeva, E. V Vorontsov, *Russ. Chem. Bull.* **1997**, *46*, 1897–1900. *Synthesis of [2.2]paracyclophane-pseudo-ortho-dicarboxylic acid*.

[158] B. Jiang, X.-L. Zhao. and X.-Y. Xu, *Tetrahedron: Asymmetry* **2005**, *16*, 1071–1074. *Resolution of (±)-[2.2]paracyclophane-4,12-dicarboxylic acid.*

[159] H. J. Reich, D. J. Cram, *J. Am. Chem. Soc.* **1969**, *91*, 3527–3533. *Macro Rings. XXXVII. Multiple Electrophilic Substitution Reactions of [2.2]Paracyclophanes and Interconversions of Polysubstituted Derivatives.*

[160] D. Enders, M. Ludwig, G. Raabe, *Chirality* **2012**, *24*, 215–222. *Synthesis and Application of the First Planar Chiral Strong Brønsted Acid Organocatalysts.*

[161] I. Piel, J. V. Dickschat, T. Pape, F. E. Hahn, F. Glorius, *Dalt. Trans.* **2012**, *41*, 13788. *A planar chiral [2.2]paracyclophane derived N-heterocyclic stannylene.*

[162] H. J. Reich, D. J. Cram, *J. Am. Chem. Soc.* **1969**, *91*, 3517–3526. *Macro Rings. XXXVI. Ring Expansion, Racemization, and Isomer Interconversions in the [2.2]Paracyclophane System through a Diradical Intermediate.*

[163] D. J. Cram, H. P. Fischer, *J. Org. Chem.* **1965**, *30*, 1815–1819. *Macro Rings. XXX. Structure of Anomalous Products of Acylation of [2.2]Paracyclophane.*

[164] H. Vorbrüggen, K. Krolikiewicz, *Tetrahedron Lett.* **1981**, *22*, 4471–4474. *A simple synthesis of Δ^2-oxazolines, Δ^2-oxazines, Δ^2-thiazolines and Δ^2-imidazoline.*

[165] A. Chesney, M. R. Bryce, *Tetrahedron Asymmetry* **1996**, *7*, 3247–3254. *Chiral oxazolines linked to tetrathiafulvalene (TTF): Redox-active ligands for asymmetric synthesis.*

[166] D. L. Davies, O. Al-Duaij, J. Fawcett, K. Singh, *Organometallics* **2010**, *29*, 1413–1420. *Reactions of cyclometalated oxazoline half-sandwich complexes of iridium and ruthenium with alkynes and CO.*

[167] Y. Boutadla, D. L. Davies, R. C. Jones, K. Singh, *Chem. - A Eur. J.* **2011**, *17*, 3438–3448. *The scope of ambiphilic acetate-assisted cyclometallation with half-sandwich complexes of iridium, rhodium and ruthenium.*

[168] W.-G. Jia, T. Zhang, D. Xie, Q.-T. Xu, S. Ling, Q. Zhang, *Dalt. Trans.* **2016**, *45*, 14230–14237. *Half-sandwich cycloruthenated complexes from aryloxazolines: synthesis, structures, and catalytic activities.*

[169] B. Li, T. Roisnel, C. Darcel, P. H. Dixneuf, *Dalt. Trans.* **2012**, *41*, 10934. *Cyclometallation of arylimines and nitrogen-containing heterocycles via room-temperature C–H bond activation with arene ruthenium(II) acetate complexes.*

[170] A. D. Ryabov, V. S. Sukharev, L. Alexandrova, R. Le Lagadec, M. Pfeffer, *Inorg. Chem.* **2001**, *40*, 6529–6532. *New synthesis and new bio-application of cyclometalated ruthenium(II) complexes for fast mediated electron transfer with peroxidase and glucose oxidase.*

[171] B. Boff, M. Ali, L. Alexandrova, N. Á. Espinosa-Jalapa, R. O. Saavedra-Díaz, R. Le Lagadec, M. Pfeffer, *Organometallics* **2013**, *32*, 5092–5097. *Rational synthesis of*

heteroleptic tris(chelate) ruthenium complexes [RuII(2-Ph-2'-Py)(L̂L)(L'L̂')]PF6 by selective substitution of the ligand trans to the ruthenated phenyl ring.

[172]   V. Martínez Cornejo, J. Olvera Mancilla, S. López Morales, J. A. Oviedo Fortino, S. Hernández-Ortega, L. Alexandrova, R. Le Lagadec, *J. Organomet. Chem.* **2015**, *799–800*, 299–310. *Synthesis and comparative behavior of ruthena(II)cycles bearing benzene ligand in the radical polymerization of styrene and vinyl acetate.*

[173]   C. Braun, M. Nieger, W. R. Thiel, S. Bräse, *Chem. - A Eur. J.* **2017**, *23*, 15474–15483. *[2.2]Paracyclophanes with N-Heterocycles as Ligands for Mono- and Dinuclear Ruthenium(II) Complexes.*

[174]   A. Pelter, H. Kidwell, R. A. N. C. Crump, *J. Chem. Soc., Perkin Trans.* **1997**, 3137–3139. *N-Methyl- and N-benzyl-4-amino [2.2] paracyclophanes as unique planar chiral auxiliaries.*

[175]   V. V. Kane, A. Gerdes, W. Grahn, L. Ernst, I. Dix, P. G. Jones, H. Hopf, *Tetrahedron Lett.* **2001**, *42*, 373–376. *A novel entry into a new class of cyclophane derivatives: Synthesis of (±)-[2.2]paracyclophane-4-thiol.*

[176]   V. I. Rozenberg, N. V Dubrovina, E. V Vorontsov, E. V Sergeeva, Y. N. Belokon, *Tetrahedron: Asymmetry* **1999**, *10*, 511–517. *New chiral β-diketones based on [2.2] paracyclophane.*

[177]   C. Bolm, K. Wenz, G. Raabe, *J. Organomet. Chem.* **2002**, *662*, 23–33. *Regioselective palladation of 2-oxazolinyl-[2.2]paracyclophanes. Synthesis of planar-chiral phosphines.*

[178]   S. Chanthamath, H. S. A. Mandour, T. M. T. Tong, K. Shibatomi, S. Iwasa, *Chem. Commun.* **2016**, *52*, 7814–7817. *Highly stereoselective cyclopropanation of diazo Weinreb amides catalyzed by chiral Ru(II)–Amm–Pheox complexes.*

[179]   S. Chanthamath, S. Iwasa, *Acc. Chem. Res.* **2016**, *49*, 2080–2090. *Enantioselective Cyclopropanation of a Wide Variety of Olefins Catalyzed by Ru(II)-Pheox Complexes.*

[180]   S. Chanthamath, S. Takaki, K. Shibatomi, S. Iwasa, *Angew. Chemie - Int. Ed.* **2013**, *52*, 5818–5821. *Highly stereoselective cyclopropanation of α,β-unsaturated carbonyl compounds with methyl (diazoacetoxy)acetate catalyzed by a chiral ruthenium(II) complex.*

[181]   Y. Nakagawa, S. Chanthamath, K. Shibatomi, S. Iwasa, *Org. Lett.* **2015**, *17*, 2792–2795. *Ru(II)-Pheox-catalyzed asymmetric intramolecular cyclopropanation of electron-deficient olefins.*

[182]   S. Chanthamath, K. Phomkeona, K. Shibatomi, S. Iwasa, *Chem. Commun.* **2012**, *48*, 7750. *Highly stereoselective Ru(II)–Pheox catalyzed asymmetric cyclopropanation of terminal olefins with succinimidyl diazoacetate.*

[183]   S. Chanthamath, D. T. Nguyen, K. Shibatomi, S. Iwasa, *Org. Lett.* **2013**, *15*, 772–775. *Highly enantioselective synthesis of cyclopropylamine derivatives via Ru(II)-pheox-catalyzed direct asymmetric cyclopropanation of vinylcarbamates.*

[184]   S. Chanthamath, S. Ozaki, K. Shibatomi, S. Iwasa, *Org. Lett.* **2014**, *16*, 3012–3015. *Highly stereoselective synthesis of cyclopropylphosphonates catalyzed by chiral Ru(II)-pheox complex.*

[185]   Y. Nakagawa, S. Chanthamath, I. Fujisawa, K. Shibatomi, S. Iwasa, *Chem. Commun.* **2017**, *53*, 3753–3756. *Ru(II)-Pheox-catalyzed Si–H insertion reaction: construction of enantioenriched carbon and silicon centers.*

[186]   A. A. Elfotoh, H. W. Chua, S. Murakami, K. Phomkeona, K. Shibatomi, *Adv. Mater. Res.* **2013**, *626*, 411–414. *A Novel Porous-Polymer-Supported Ru(II)-dm-Pheox Catalyst and its Application in Highly Efficient N-H Insertion Reactions.*

[187]   E. Ferrer Flegeau, C. Bruneau, P. H. Dixneuf, A. Jutand, *J. Am. Chem. Soc.* **2011**, *133*, 10161–10170. *Autocatalysis for C–H Bond Activation by Ruthenium(II) Complexes in Catalytic Arylation of Functional Arenes.*

[188]   D. W. Old, J. P. Wolfe, S. L. Buchwald, *J. Am. Chem. Soc.* **1998**, *120*, 9722–9723. *A highly active catalyst for palladium-catalyzed cross-coupling reactions: room-temperature Suzuki couplings and amination of unactivated aryl chlorides.*

[189]   J. M. Fox, X. Huang, A. Chieffi, S. L. Buchwald, *J. Am. Chem. Soc.* **2000**, *122*, 1360–1370. *Highly active and selective catalysts for the formation of α-aryl ketones.*

[190]   K. L. Billingsley, T. E. Barder, S. L. Buchwald, *Angew. Chemie - Int. Ed.* **2007**, *46*, 5359–5363. *Palladium-catalyzed borylation of aryl chlorides: Scope, applications, and computational studies.*

[191]   K. L. Billingsley, S. L. Buchwald, *J. Org. Chem.* **2008**, *73*, 5589–5591. *An improved system for the palladium-catalyzed borylation of aryl halides with pinacol borane.*

[192]   R. Martin, S. L. Buchwald, *J. Am. Chem. Soc.* **2007**, *129*, 3844–3845. *Pd-catalyzed Kumada-Corriu cross-coupling reactions at low temperatures allow the use of Knochel-type Grignard reagents.*

[193]   E. McNeill, T. E. Barder, S. L. Buchwald, *Org. Lett.* **2007**, *9*, 3785–3788. *Palladium-catalyzed silylation of aryl chlorides with hexamethyldisilane.*

[194]   M. Dračínský, P. Jansa, P. Bouř, *Chem. - A Eur. J.* **2012**, *18*, 981–986. *Computational and experimental evidence of through-space NMR spectroscopic J coupling of hydrogen atoms.*

[195]   Y. Wang, Y. Zhu, Z. Chen, A. Mi, W. Hu, M. P. Doyle, *Org. Lett.* **2003**, *5*, 3923–3926. *A Novel Three-Component Reaction Catalyzed by Dirhodium(II) Acetate: Decomposition of Phenyldiazoacetate with Arylamine and Imine for Highly Diastereoselective Synthesis of 1,2-Diamines.*

[196]   H. Huang, Y. Wang, Z. Chen, W. Hu, *Adv. Synth. Catal.* **2005**, *347*, 531–534. *Rhodium-catalyzed, three-component reaction of diazo compounds with amines and azodicarboxylates.*

[197]   A. Ceylan, S. and Kirschning, in *Recover. Recycl. Catal.*, John Wiley & Sons, Ltd, Chichester, UK, **2009**.

[198]   C. Wiles, P. Watts, *European J. Org. Chem.* **2008**, 1655–1671. *Continuous flow reactors, a tool for the modern synthetic chemist.*

[199]   X. Y. Mak, P. Laurino, P. H. Seeberger, *Beilstein J. Org. Chem.* **2009**, *5*, 1–11. *Asymmetric reactions in continuous flow.*

[200]   C. W. Bradshaw, C. H. Wong, W. Hummel, *J. Org. Chem.* **1992**, *57*, 1532–1536. *Lactobacillus kefir Alcohol Dehydrogenase: A Useful Catalyst for Synthesis.*

[201]   S. Rodríguez, K. T. Schroeder, M. M. Kayser, J. D. Stewart, *J. Org. Chem.* **2000**, *65*, 2586–2587. *Asymmetric synthesis of β-hydroxy esters and α-alkyl-β-hydroxy esters by recombinant Escherichia coli expressing enzymes from baker's yeast.*

[202]   G. Ford, E. M. Ellis, *Yeast* **2002**, *19*, 1087–1096. *Characterization of Ypr1p from Saccharomyces cerevisiae as a 2-methylbutyraldehyde reductase.*

[203]   M. Muller, M. Katzberg, M. Bertau, W. Hummel, *Org. Biomol. Chem.* **2010**, *8*, 1540–1550. *Highly efficient and stereoselective biosynthesis of (2S,5S)-hexanediol with a dehydrogenase from Saccharomyces cerevisiae.*

[204]   J. Cossy, A. Guérinot, in *Heterocycl. Chem. 21st Century* (Eds.: E. Scriven, C. Ramsden), Academic Press, **2016**, pp. 107–142.

[205]   B. M. Trost, A. C. Gutierrez, R. C. Livingston, *Org. Lett.* **2009**, *11*, 2539–2542. *Tandem Ruthenium-Catalyzed Redox Isomerization-O-Conjugate Addition: An Atom-Economic Synthesis of Cyclic Ethers.*

[206]   I. Vilotijevic, T. F. Jamison, *Science.* **2007**, *317*, 1189-1192. *Epoxide-Opening Cascades Promoted by Water.*

[207]   T. Bach, K. Kather, *J. Org. Chem.* **1996**, *61*, 7642–7643. *Intramolecular Nucleophilic Substitution at the C-4 Position of Functionalized Oxetanes: A Ring Expansion for the Construction of Various Heterocycles.*

[208]   F. Della-Felice, F. de Assis, A. Sarotti, R. Pilli, *Synthesis (Stuttg).* **2019**, DOI 10.1055/s-0037-1611708 *Palladium-Catalyzed Formation of Substituted Tetrahydropyrans: Mechanistic Insights and Structural Revision of Natural Products.*

[209]   B. Godoi, R. F. Schumacher, G. Zeni, *Chem. Rev.* **2011**, *111*, 2937–2980. *Synthesis of Heterocycles via Electrophilic Cyclization of Alkynes Containing Heteroatom.*

[210]   X.-G. Song, S.-F. Zhu, X.-L. Xie, Q.-L. Zhou, *Angew. Chemie* **2013**, *125*, 2615–2618. *Enantioselective Copper-Catalyzed Intramolecular Phenolic O-H nucleophilic ring-opening of epoxidesnucleophilic ring-opening of epoxidesH Bond Insertion: Synthesis of Chiral 2-Carboxy Dihydrobenzofurans, Dihydrobenzopyrans, and Tetrahydrobenzooxepines.*

[211]   M. P. Moyer, P. L. Feldman, H. Rapoport, *J. Org. Chem.* **1985**, *50*, 5223–5230. *Intramolecular nitrogen-hydrogen, oxygen-hydrogen and sulfur-hydrogen insertion reactions. Synthesis of heterocycles from α-diazo β-keto esters.*

[212]   J. Hansen, B. Li, E. Dikarev, J. Autschbach, H. M. L. Davies, *J. Org. Chem.* **2009**, *74*, 6564–6571. *Combined experimental and computational studies of heterobimetallic Bi-Rh paddlewheel carboxylates as catalysts for metal carbenoid transformations.*

[213]   J. Hansen, H. M. L. Davies, *Coord. Chem. Rev.* **2008**, *252*, 545–555. *High symmetry dirhodium(II) paddlewheel complexes as chiral catalysts.*

[214]   H. M. L. Davies, S. A. Panaro, *Tetrahedron Lett.* **1999**, *40*, 5287–5290. *Novel dirhodium tetraprolinate catalysts containing bridging prolinate ligands for asymmetric carbenoid reactions.*

[215]   H. M. L. Davies, B. Hu, *Tetrahedron Lett.* **1992**, *33*, 455–456. *Regioselective [3+2] annulations with rhodium(II)-stabilized vinylcarbenoids.*

[216]   H. M. L. Davies, P. R. Bruzinski, D. H. Lake, N. Kong, M. J. Fall, *J. Am. Chem. Soc.* **1996**, *118*, 6897–6907. *Asymmetric Cyclopropanations by Rhodium(II) N-(Arylsulfonyl)prolinate Catalyzed Decomposition of Vinyldiazomethanes in the Presence of Alkenes. Practical Enantioselective Synthesis of the Four Stereoisomers of 2-Phenylcyclopropan-1-amino Acid.*

[217]   R. P. Reddy, G. H. Lee, H. M. L. Davies, *Org. Lett.* **2006**, *8*, 3437–3440. *Dirhodium Tetracarboxylate Derived from Adamantylglycine as a Chiral Catalyst for Carbenoid Reactions.*

[218]   J. R. Denton, D. Sukumaran, H. M. L. Davies, *Org. Lett.* **2007**, *9*, 2625–2628. *Enantioselective Synthesis of Trifluoromethyl-Substituted Cyclopropanes.*

[219]   C. Liang, F. Robert-Peillard, C. Fruit, P. Müller, R. H. Dodd, P. Dauban, *Angew. Chemie Int. Ed.* **2006**, *45*, 4641–4644. *Efficient Diastereoselective Intermolecular Rhodium-Catalyzed C–H Amination.*

[220]   H. M. L. Davies, R. J. Townsend, *J. Org. Chem.* **2001**, *66*, 6595–6603. *Catalytic Asymmetric Cyclopropanation of Heteroaryldiazoacetates.*

[221]   H. M. L. Davies, N. Kong, M. R. Churchill, *J. Org. Chem.* **1998**, *63*, 6586–6589. *Asymmetric Synthesis of Cyclopentenes by [3+2] Annulations between Vinylcarbenoids and Vinyl Ethers.*

[222]   IUPAC, **1998** *Phane Nomenclature - Part I: Phane Parents Names.*

[223]   F. Vögtle, *Tetrahedron Lett.* **1969**, *121*, 5329–5334. *Zur Nomenklatur der Phane.*

[224]   R. S. Cahn, C. Ingold, V. Prelog, *Angew. Chemie Int. Ed. English* **1966**, *5*, 385–415. *Specification of Molecular Chirality.*

[225]   L. Ernst, *Liebigs Ann.* **1994**, *1995*, 13–17. *The conformational equilibrium of [2.2]paracyclophanes in solution.*

[226] D. J. Cram, A. C. Day, *J. Org. Chem.* **1966**, *31*, 1227–1232. *Macro Rings. XXXI. Quinone Derived from [2.2]Paracyclophane, an Intramolecular-Molecular Complex.*

[227] L. Ernst, L. Wittkowski, *European J. Org. Chem.* **1999**, 1653–1663. *Diastereomers Composed of Two Planar-Chiral Subunits : Bis ([2.2] paracyclophan-4-yl) methane and Analogues.*

[228] L. Bondarenko, I. Dix, H. Hinrichs, H. Hopf, *Synthesis (Stuttg).* **2004**, 2751–2759. *Cyclophanes. Part LII: Ethynyl[2.2]paracyclophanes - New building blocks for molecular scaffolding.*

[229] D. C. Braddock, I. D. MacGilp, B. G. Perry, *J. Org. Chem.* **2002**, *67*, 8679–8681. *Improved synthesis of (±)-4,12-dihydroxy[2.2]paracyclophane and Its enantiomeric resolution by enzymatic methods: Planar chiral (R)- and (S)-phanol.*

[230] J. F. Schneider, F. C. Falk, R. Fröhlich, J. Paradies, *European J. Org. Chem.* **2010**, 2265–2269. *Planar-chiral thioureas as hydrogen-bond catalysts.*

[231] C. Braun, *Synthese Und Anwendung Planar-Chiraler [2.2]Paracyclophane Mit N-Donorfunktionen Und Deren Übergangsmetallkomplexe Zur Synthese Substituierter Amine*, Karlsruhe Institut für Technologie, **2017**.

[232] K. P. Jayasundera, T. G. W. Engels, D. J. Lun, M. N. Mungalpara, P. G. Plieger, G. J. Rowlands, *Org. Biomol. Chem.* **2017**, *15*, 8975–8984. *The synthesis of planar chiral pseudo-gem aminophosphine pre-ligands based on [2.2]paracyclophane.*

[233] H. Allgeier, M. G. Siegel, R. C. Helgeson, E. Schmidt, D. Cram, *J. Am. Chem. Soc.* **1975**, *97*, 3782–3789. *Macro Rings. XLVII. Syntheses and Spectral Properties of Heteroannularly Disubstituted [2.2]Paracyclophanes.*

[234] S. Sugiyama, Y. Aoki, K. Ishii, *Tetrahedron Asymmetry* **2006**, *17*, 2847–2856. *Enantioselective addition of diethylzinc to aldehydes catalyzed by monosubstituted [2.2]paracyclophane-based N,O-ligands: remarkable cooperative effects of planar and central chiralities.*

[235] J. L. Marshall, L. Hall, *Tetrahedron* **1981**, *37*, 1271–1275. *THE ELECTRONIC SPECTRUM OF (-)-S-(pS)-2,5,3',6'-TETRAHYDR0[2.2]PARACYCLOPHANE-2-CARBOXYLIC ACID.*

[236] P. B. Hitchcock, G. J. Rowlands, R. Parmar, *Chem. Commun.* **2005**, 4219–4221. *The synthesis of enantiomerically pure 4-substituted [2.2]paracyclophane derivatives by sulfoxide-metal exchange.*

[237] M. Kreis, C. J. Friedmann, S. Bräse, *Chem. - A Eur. J.* **2005**, *11*, 7387–7394. *Diastereoselective Hartwig-Buchwald reaction of chiral amines with rac-[2.2]paracyclophane derivatives.*

[238] J. E. Glover, D. J. Martin, P. G. Plieger, G. J. Rowlands, *European J. Org. Chem.* **2013**, *3*, 1671–1675. *Planar chiral triazole-based phosphanes derived from [2.2]paracyclophane and their activity in suzuki coupling reactions.*

[239]   C. Braun, S. Bräse, L. L. Schafer, *European J. Org. Chem.* **2017**, *2017*, 1760–1764. *Planar-Chiral [2.2]Paracyclophane-Based Amides as Proligands for Titanium- and Zirconium-Catalyzed Hydroamination.*

[240]   K. P. Jayasundera, D. N. M. Kusmus, L. Deuilhé, L. Etheridge, Z. Farrow, D. J. Lun, G. Kaur, G. J. Rowlands, *Org. Biomol. Chem.* **2016**, *14*, 10848–10860. *The synthesis of substituted amino[2.2]paracyclophanes.*

[241]   D. K. Whelligan, C. Bolm, *J. Org. Chem.* **2006**, *71*, 4609–4618. *Synthesis of Pseudo-geminal-, Pseudo-ortho-, and ortho-Phosphinyl-oxazolinyl-[2.2]paracyclophanes for Use as Ligands in Asymmetric Catalysis.*

[242]   J. Elaridi, A. Thaqi, A. Prosser, W. Jackson, A. Robinson, *Tetrahedron Asymmetry* **2005**, *16*, 1309–1319. *An enantioselective synthesis of β²-amino acid derivatives.*

[243]   L. C. Morrill, T. Lebl, A. M. Z. Slawin, A. D. Smith, *Chem. Sci.* **2012**, *3*, 2088–2093. *Catalytic asymmetric α-amination of carboxylic acids using isothioureas.*

[244]   P. Saha, H. Jeon, P. K. Mishra, H. W. Rhee, J. H. Kwak, *J. Mol. Catal. A Chem.* **2016**, *417*, 10–18. *N-H and S–H insertions over Cu(I)-zeolites as heterogeneous catalysts.*

[245]   K. H. Chan, X. Guan, V. K. Y. Lo, C. M. Che, *Angew. Chemie - Int. Ed.* **2014**, *53*, 2982–2987. *Elevated catalytic activity of ruthenium(II)-porphyrin-catalyzed carbene/nitrene transfer and insertion reactions with n-heterocyclic carbene ligands.*

[246]   P. Le Maux, G. Simonneaux, *Tetrahedron* **2015**, *71*, 9333–9338. *Enantioselective insertion of carbenoids into N-H bonds catalyzed by chiral bicyclobisoxazoline copper(I) complexes.*

[247]   J. Bang, J. Kim, J. Kim, C. M. Yu, *Org. Lett.* **2015**, *17*, 1573–1576. *Asymmetric aldol reaction of allenoates: Regulation for the selective formation of isomeric allenyl or alkynyl aldol adduct.*

[248]   L. Rout, A. M. Harned, *Chem. - A Eur. J.* **2009**, *15*, 12926–12928. *Allene carboxylates as dipolarophiles in Rh-catalyzed carbonyl ylide cycloadditions.*

[249]   G. Wang, X. H. Liu, Y. Chen, J. Yang, J. Li, L. Lin, X. M. Feng, *ACS Catal.* **2016**, *6*, 2482–2486. *Diastereoselective and Enantioselective Alleno-aldol Reaction of Allenoates with Isatins to Synthesis of Carbinol Allenoates Catalyzed by Gold.*

[250]   X. Hu, J. Sun, H. G. Wang, R. Manetsch, *J. Am. Chem. Soc.* **2008**, *130*, 13820–13821. *Bcl-XL-templated assembly of its own protein-protein interaction modulator from fragments decorated with thio acids and sulfonyl azides.*

[251]   K. Nozaki, K. Oshima, K. Utimoto, *Chem. Soc. Japan* **1991**, *64*, 403–409. *Trialkylborane as an Initiator and Terminator of Free Radical Reactions. Facile Routes to Borons Enolates via alpha-carbonyl radiacals and aldol reactions of boron enolates.*

[252]   J. Barluenga, F. González-Bobes, M. C. Murguía, S. R. Ananthoju, J. M. González, *Chem. - A Eur. J.* **2004**, *10*, 4206–4213. *Bis(pyridine)iodonium tetrafluoroborate (IPy2BF4): A versatile oxidizing reagent.*

[253]   C. S. A. Antunes, M. Bietti, O. Lanzalunga, M. Salamone, *J. Org. Chem.* **2004**, *69*, 5281–5289. *Photolysis of 1-alkylcycloalkanols in the presence of (diacetoxyiodo) benzene and I2. Intramolecular selectivity in the β-scission reactions of the intermediate 1-alkylcycloalkoxyl radicals.*

[254]   F. Q. Huang, J. Xie, J. G. Sun, Y. W. Wang, X. Dong, L. W. Qi, B. Zhang, *Org. Lett.* **2016**, *18*, 684–687. *Regioselective Synthesis of Carbonyl-Containing Alkyl Chlorides via Silver-Catalyzed Ring-Opening Chlorination of Cycloalkanols.*

[255]   W. Li, F. Tan, X. Hao, G. Wang, Y. Tang, X. Liu, L. Lin, X. Feng, *Angew. Chemie - Int. Ed.* **2015**, *54*, 1608–1611. *Catalytic asymmetric intramolecular homologation of ketones with α-diazoesters: Synthesis of cyclic α-Aryl/Alkyl β-ketoesters.*

9 Appendix

9.1 X-ray Crystallographic data

(*rac*)-4-nitro[2.2]paracyclophane

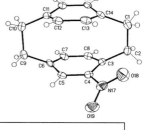

SB893_HY

Crystal data

$C_{16}H_{15}NO_2$	$F(000) = 535$
$M_r = 253.29$	$D_x = 1.365$ Mg m^{-3}
Monoclinic, Cc (no.9)	Cu $K\alpha$ radiation, $\lambda = 1.54178$ Å
$a = 14.3361$ (4) Å	Cell parameters from 5673 reflections
$b = 7.4134$ (2) Å	$\theta = 6.8–72.1°$
$c = 11.6019$ (3) Å	$\mu = 0.72$ mm^{-1}
$\beta = 91.249$ (1)°	$T = 123$ K
$V = 1232.75$ (6) Å^3	Blocks, colourless
$Z = 4$	$0.24 \times 0.14 \times 0.06$ mm

Data collection

Bruker D8 VENTURE diffractometer with Photon100 detector	2244 independent reflections
Radiation source: INCOATEC microfocus sealed tube	2178 reflections with $I > 2\sigma(I)$
Detector resolution: 10.4167 pixels mm^{-1}	$R_{int} = 0.025$
rotation in ϕ and ω, 1°, shutterless scans	$\theta_{max} = 72.2°$, $\theta_{min} = 6.2°$
Absorption correction: multi-scan *SADABS* (Sheldrick, 2014)	$h = -17 \rightarrow 16$
$T_{min} = 0.835$, $T_{max} = 0.958$	$k = -9 \rightarrow 8$
6734 measured reflections	$l = -14 \rightarrow 14$

Refinement

Refinement on F^2	Secondary atom site location: difference Fourier map
Least-squares matrix: full	Hydrogen site location: difference Fourier map
$R[F^2 > 2\sigma(F^2)] = 0.046$	H-atom parameters constrained
$wR(F^2) = 0.125$	$w = 1/[\sigma^2(F_o^2) + (0.083P)^2 + 0.7543P]$ where $P = (F_o^2 + 2F_c^2)/3$
$S = 1.08$	$(\Delta/\sigma)_{max} < 0.001$
2244 reflections	$\Delta_{max} = 0.38$ e Å^{-3}
172 parameters	$\Delta_{min} = -0.26$ e Å^{-3}
2 restraints	Absolute structure: Flack x determined using 961 quotients [(I+)-(I-)]/[(I+)+(I-)] (Parsons, Flack and Wagner, Acta Cryst. B69 (2013) 249-259). The absolute structure cannot be determined reliably.
Primary atom site location: structure-invariant direct methods	Absolute structure parameter: 0.32 (14)

(rac)-13-amino-4-bromo[2.2]paracyclophane

SB_862HY

Crystal data

$C_{16}H_{16}BrN$	$F(000) = 616$
$M_r = 302.21$	$D_x = 1.620$ Mg m^{-3}
Monoclinic, $P2_1/c$ *(no.14)*	Cu $K\alpha$ radiation, $\lambda = 1.54178$ Å
$a = 14.6067$ (7) Å	Cell parameters from 9816 reflections
$b = 7.4498$ (4) Å	$\theta = 3.9–72.1°$
$c = 11.4155$ (6) Å	$\mu = 4.33$ mm^{-1}
$\beta = 93.825$ (1)°	$T = 123$ K
$V = 1239.43$ (11) Å^3	Blocks, colourless
$Z = 4$	$0.40 \times 0.20 \times 0.08$ mm

Data collection

Bruker D8 VENTURE diffractometer with Photon100 detector	2426 independent reflections
Radiation source: INCOATEC microfocus sealed tube	2411 reflections with $I > 2\sigma(I)$
Detector resolution: 10.4167 pixels mm^{-1}	$R_{int} = 0.026$
rotation in ϕ and ω, 1°, shutterless scans	$\theta_{max} = 72.1°$, $\theta_{min} = 6.1°$
Absorption correction: multi-scan *SADABS* (Sheldrick, 2014)	$h = -15 \rightarrow 17$
$T_{min} = 0.455$, $T_{max} = 0.715$	$k = -9 \rightarrow 9$
16710 measured reflections	$l = -13 \rightarrow 14$

Refinement

Refinement on F^2	Secondary atom site location: difference Fourier map
Least-squares matrix: full	Hydrogen site location: difference Fourier map
$R[F^2 > 2\sigma(F^2)] = 0.026$	H atoms treated by a mixture of independent and constrained refinement
$wR(F^2) = 0.064$	$w = 1/[\sigma^2(F_o^2) + (0.0214P)^2 + 2.057P]$ where $P = (F_o^2 + 2F_c^2)/3$
$S = 1.11$	$(\Delta/\sigma)_{max} = 0.001$
2426 reflections	$\Delta\rangle_{max} = 0.66$ e Å^{-3}
170 parameters	$\Delta\rangle_{min} = -0.30$ e Å^{-3}
2 restraints	Extinction correction: *SHELXL2014/7* (Sheldrick, 2014), Fc*=kFc[1+0.001xFc$^2\lambda^3$/sin(2θ)]$^{-1/4}$
Primary atom site location: structure-invariant direct methods	Extinction coefficient: 0.00190 (17)

(*rac*)-13,13-dimethylamino-4-bromo[2.2]paracyclophane

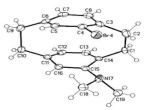

SB891_HY

Crystal data

$C_{18}H_{20}BrN$	$D_x = 1.519$ Mg m^{-3}
$M_r = 330.26$	Cu $K\alpha$ radiation, $\lambda = 1.54178$ Å
Orthorhombic, $Pna2_1$ (no.33)	Cell parameters from 6200 reflections
$a = 16.9814$ (6) Å	$\theta = 5.2$–72.1°
$b = 7.6914$ (3) Å	$\mu = 3.77$ mm^{-1}
$c = 11.0584$ (4) Å	$T = 123$ K
$V = 1444.35$ (9) Å^3	Plates, colourless
$Z = 4$	0.20 × 0.18 × 0.02 mm
$F(000) = 680$	

Data collection

Bruker D8 VENTURE diffractometer with Photon100 detector	2613 independent reflections
Radiation source: INCOATEC microfocus sealed tube	2382 reflections with $I > 2\sigma(I)$
Detector resolution: 10.4167 pixels mm^{-1}	$R_{int} = 0.041$
rotation in ϕ and ω, 1°, shutterless scans	$\theta_{max} = 72.3°$, $\theta_{min} = 5.2°$
Absorption correction: multi-scan $SADABS$ (Sheldrick, 2014)	$h = -20 \rightarrow 20$
$T_{min} = 0.704$, $T_{max} = 0.915$	$k = -9 \rightarrow 9$
9451 measured reflections	$l = -13 \rightarrow 12$

Refinement

Refinement on F^2	Secondary atom site location: difference Fourier map
Least-squares matrix: full	Hydrogen site location: difference Fourier map
$R[F^2 > 2\sigma(F^2)] = 0.046$	H-atom parameters constrained
$wR(F^2) = 0.120$	$w = 1/[\sigma^2(F_o^2) + (0.076P)^2 + 1.439P]$ where $P = (F_o^2 + 2F_c^2)/3$
$S = 1.04$	$(\Delta/\sigma)_{max} < 0.001$
2613 reflections	$\Delta\rangle_{max} = 0.68$ e Å^{-3}
183 parameters	$\Delta\rangle_{min} = -0.76$ e Å^{-3}
1 restraint	Absolute structure: Flack x determined using 896 quotients [(I+)-(I-)]/[(I+)+(I-)] (Parsons, Flack and Wagner, Acta Cryst. B69 (2013) 249-259).
Primary atom site location: dual	Absolute structure parameter: -0.07 (4)

Methyl 2-phenyl-2-(phenylamino)acetate

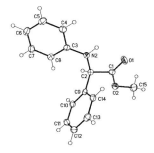

SB826_HY

Crystal data

$C_{15}H_{15}NO_2$	$F(000) = 1024$
$M_r = 241.28$	$D_x = 1.280$ Mg m^{-3}
Monoclinic, $C2/c$ *(no.15)*	Cu $K\alpha$ radiation, $\lambda = 1.54178$ Å
$a = 18.1610\,(5)$ Å	Cell parameters from 6322 reflections
$b = 5.4218\,(1)$ Å	$\theta = 3.5$–$72.1°$
$c = 25.4396\,(7)$ Å	$\mu = 0.68$ mm^{-1}
$\beta = 91.833\,(1)°$	$T = 123$ K
$V = 2503.64\,(11)$ Å^3	Plates, colourless
$Z = 8$	$0.18 \times 0.10 \times 0.06$ mm

Data collection

Bruker D8 Venture diffractometer with Photon100 detector	2200 reflections with $I > 2\sigma(I)$
Radiation source: IμS microfocus	$R_{int} = 0.027$
rotation in ϕ and ω, 1°, shutterless scans	$\theta_{max} = 72.1°$, $\theta_{min} = 3.5°$
Absorption correction: multi-scan *SADABS* (Sheldrick, 2015)	$h = -22 \rightarrow 22$
$T_{min} = 0.891$, $T_{max} = 0.958$	$k = -6 \rightarrow 6$
9009 measured reflections	$l = -30 \rightarrow 31$
2449 independent reflections	

Refinement

Refinement on F^2	Secondary atom site location: difference Fourier map
Least-squares matrix: full	Hydrogen site location: difference Fourier map
$R[F^2 > 2\sigma(F^2)] = 0.032$	H atoms treated by a mixture of independent and constrained refinement
$wR(F^2) = 0.080$	$w = 1/[\sigma^2(F_o^2) + (0.0273P)^2 + 1.8913P]$ where $P = (F_o^2 + 2F_c^2)/3$
$S = 1.07$	$(\Delta/\sigma)_{max} < 0.001$
2449 reflections	$\Delta\rangle_{max} = 0.21$ e Å^{-3}
168 parameters	$\Delta\rangle_{min} = -0.15$ e Å^{-3}
1 restraint	Extinction correction: *SHELXL*, Fc*=kFc[1+0.001xFc$^2\lambda^3$/sin(2θ)]$^{-1/4}$
Primary atom site location: structure-invariant direct methods	Extinction coefficient: 0.00130 (10)

6-(*tert*-butyl) 1-methyl (*E*)-5-(phenylamino)-hex-2-enedioate

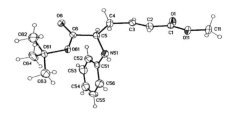

SB968_HY

Crystal data

$C_{17}H_{23}NO_4$	$F(000) = 656$
$M_r = 305.36$	$D_x = 1.203$ Mg m^{-3}
Monoclinic, $P2_1/n$ *(no.14)*	Cu $K\alpha$ radiation, $\lambda = 1.54178$ Å
$a = 5.8352$ (1) Å	Cell parameters from 9954 reflections
$b = 18.4613$ (5) Å	$\theta = 3.7–72.1°$
$c = 15.8759$ (4) Å	$\mu = 0.70$ mm^{-1}
$\beta = 99.728$ (1)°	$T = 123$ K
$V = 1685.65$ (7) Å^3	Needle, colourless
$Z = 4$	$0.27 \times 0.06 \times 0.03$ mm

Data collection

Bruker D8 VENTURE diffractometer with Photon100 detector	3318 independent reflections
Radiation source: INCOATEC microfocus sealed tube	3114 reflections with $I > 2\sigma(I)$
Detector resolution: 10.4167 pixels mm^{-1}	$R_{int} = 0.029$
rotation in ϕ and ω, 1°, shutterless scans	$\theta_{max} = 72.2°$, $\theta_{min} = 3.7°$
Absorption correction: multi-scan *SADABS* (Sheldrick, 2014)	$h = -7 \rightarrow 7$
$T_{min} = 0.867$, $T_{max} = 0.971$	$k = -22 \rightarrow 22$
17085 measured reflections	$l = -19 \rightarrow 19$

Refinement

Refinement on F^2	Primary atom site location: structure-invariant direct methods
Least-squares matrix: full	Secondary atom site location: difference Fourier map
$R[F^2 > 2\sigma(F^2)] = 0.064$	Hydrogen site location: mixed
$wR(F^2) = 0.150$	H atoms treated by a mixture of independent and constrained refinement
$S = 1.11$	$w = 1/[\sigma^2(F_o^2) + (0.0469P)^2 + 2.3123P]$ where $P = (F_o^2 + 2F_c^2)/3$
3318 reflections	$(\Delta/\sigma)_{max} < 0.001$
203 parameters	$\Delta\rangle_{max} = 0.91$ e Å^{-3}
1 restraint	$\Delta\rangle_{min} = -0.31$ e Å^{-3}

9.2 Curriculum Vitae

Personal Details

Name	Yuling Hu
Address	Adlerstraße 3a
	76133 Karlsruhe
	Germany
Telephone	+49 176 317 52776
E-mail	huyulinghopeful@gmail.com
Date of Birth	15.03.1990
Nationality	Chinese

Education

09.2014 – 04.2018	PhD Student in Organic Chemistry, Karlsruhe Institute of Technology, KA, Germany Thesis advisor: Prof. Dr. Stefan Bräse Thesis: *Metal-Catalyzed N–H and O–H Insertion from α-Diazocarbonyl Compounds*
09.2011 – 06.2014	Master of Science in Pharmacy Shenyang Pharmaceutical University, Liaoning, China Shanghai Institute of Organic Chemistry, Shanghai, China
	Thesis Advisor: Prof. Dr. Wenjun Tang, Assoc. Prof. Yu Sha Thesis: *A Concise Total Synthesis of Salvianolic Acid A and Enantioselective Rhodium-catalyzed Addition of Arylboronic Acid to Trifluoromethyl Ketones*
09.2007 06.2011	Bachelor of Science in Applied Chemistry Shenyang Pharmaceutical University, Liaoning, China Thesis Advisor: , Assoc. Prof. Yu Sha Thesis: *Synthesis of Imatinib-derivatives as anti-cancer Candidates*

Scholarships/Awards

05.2014	Scholarship from China Scholarship Council
02.2016	Scholarship for HeKKSaGOn winter school in Kyoto, Japan

9.3   Publications and Conference Contributions

Publications

R.S. Luo, K. Li, Y. Hu, W. J. Tang, Adv. Synth. Catal. 2013, 355, 1927-1302. Enantioselective Rhodium-catalyzed Addition of Arylboronic Acid to Trifluoromethy Ketones.

Wezeman, Y. Hu, J. McMurtrie, S. Bräse, K.-S. Masters, Austr. J. Chem. 2015, 68, 1859-1865. Synthesis of Non-Symmetrical & Axially-Chiral Dibenzo[1,3]diazepines: Pd/CPhos-Catalysed Direct Arylation of 1-(ortho-haloaryl)-3-aryl-aminals.

Y. Hu, M. Nieger, S. Bräse, in preparation. A Well-Defined Paracyclophane Catalyst for Enantioselective N–H Insertion in synthesizing Unsaturated Homo Glutamic Acid Analogues.

Y. Hu, E. Mittmann, C. Niemeyer, S. Bräse, in preparation. Enzyme Catalyzed Highly Enantioselective Reduction of α-Diazoesters: A Chemoenzymatic Synthesis of Heterocycles.

Posters

Y. Hu, S. Bräse, 27th European Colloquium of Heterocyclic Chemistry, Amsterdam, Netherlands, 03.07–06.07.2016. *A Well-Defined [2.2]Paracyclophane Copper Catalyst for Enantioselective N–H Insertion in Total Synthesis of Rostratin A–D.*

Y. Hu, S. Bräse, 26th International Society of Heterocyclic Chemistry, Regensburg, Germany, 03.09–07.09.2017. *α-Diazoesters as Key Intermediates to Access Heterocyclic Compounds.*

Oral presentation

Y. Hu, S. Bräse, HeKKSaGOn winter school, Kyoto, Japan, Feb. 2016. *Total Synthesis of Rostratin A–D.*

9.4 Acknowledgement

Firstly, I would like to express my sincere gratitude to my advisor Prof. Dr. STEFAN BRÄSE for the continuous support of my Ph.D. study and the related research, for his patience, motivation, and immense knowledge. His guidance helped me in all the time of research and writing of this thesis.

Besides my advisor, I would like to thank Prof. Dr. JOACHIM PODLECH for the acceptance of the co-reference of my thesis. All of your corrections and ideas have given me a lot of help in making this work better.

Another big thank you goes to my scholarship, *Chinese Scholarship Council* (CSC), without your continuously sponsor in the past three years, this work is even not possible to get started.

I would also like to express my gratitude to everybody at the Karlsruhe Institute of Technology who has helped me during my research, especially the department of analytics, namely Frau Kernert, Frau Ohmer and Frau Lang. Without their help, most of the data from the experimental part would not have been possible.

My cooperation partner, Esther Mittmann and Theo Peschke, Thank you very much for the support and the excellent results from the enzymatic synthesis part. Esther, you are an excellent researcher, thank you for your quick work and the beautiful figures you made for our project.

Furthermore, I would like to thank Robin Bär and Janina Beck for taking their time to proof read my work and give me very helpful suggestions and remarks until the last minute. I know that both of you have more urgent things to do. Except colleagues, both of you are good friends in my personal life. I will forever remember the "short coffee time" at 8:30 to start a normal chemistry life day and the time to discuss unsolvable questions together.

I would like to give my special thanks to our secretaries, Selin Samur and Christiane Lampert. Your organization and coordination make this group always good. Selin, as a "glückes Kind ", I wish the sweet smiles always stay on your face.

Thanks to Anne Schneider for all of your enthusiasm to help me to solve a lot of annoying problems from life and all your patience to improve my German. It is not easy to write down everything here, but all of your help will be remembered in my heart forever.

Another special thanks go to Janina Beck/Tobias Fischer and Alexander Braun/Laura Burgmaier, for giving me a small family in Karlsruhe. I will never forget the happy time as a member at your home. Alex, you are wonderful, you have your own idea and I believe you will definitely make something different. Janina, it is also not easy to write down all of my thanks here, your friendship, your warm heart as well as your suggestions make the last two years of my PhD easier.

The international lab 306, where I have worked for nearly 4 years, I would like to thank you as well as all the colleagues who have worked here, especially Dr. Larissa Geiger, Janina Beck, Robin Bär and Alena Kalyakina.

My gratitude also goes to everyone who was not mentioned before, but still took his or her time to discuss with me and help me with my chemistry: Dr. Claudia Bizarri, Dr. Zbigniew Pianowski, Dr. Carolin Braun, Eduard Spuling and Daniel Knoll.

At the end of my PhD time, I would also like to thank all the people who helped me a lot when I started my life and work in this country. Thanks to Dr. Sabilla Zhong, Dr. Stefanie Lindner, Dr. Paul Sauter.

Of course, the biggest thank you goes to my parents for their support throughout my studies. I have never been around at Chinese New Year in these four years and this will finally be over.

衷心感谢我的父母，感谢他们开明的思想，无私的帮助以及一如既往的信任。对儿女的付出也许是天下投入产出比最低的投资，然而每一份恩情女儿都会铭记在心。未来的日子里，我会承担更多的责任，与你们相知，相伴，相行。

Last but not least, I want to give my sincere thank you to my boyfriend. Thanks for your understanding, your phone call at 5:30 for 365 days as well as your "chicken soup for the soul" to help me get rid of all the stressful situations. Although there is a long geographic distance between us, our hearts are from time to time together.